CLOUD WARRIORS

CLOUD WARRIORS

DEADLY STORMS, CLIMATE CHAOS—
AND THE PIONEERS CREATING A REVOLUTION
IN WEATHER FORECASTING

THOMAS E. WEBER

ST. MARTIN'S PRESS

NEW YORK

First published in the United States by St. Martin's Press,
an imprint of St. Martin's Publishing Group

CLOUD WARRIORS. Copyright © 2025 by Thomas E. Weber. All rights reserved.
Printed in the United States of America. For information,
address St. Martin's Publishing Group, 120 Broadway, New York, NY 10271.

Portions of the text adapted from *Time* © 2018, 2019 TIME USA LLC. All rights reserved.
Used with permission.

www.stmartins.com

Designed by Steven Seighman

The Library of Congress Cataloging-in-Publication Data is available upon request.

ISBN 978-1-250-28054-1 (hardcover)
ISBN 978-1-250-28055-8 (ebook)

Our books may be purchased in bulk for promotional, educational, or
business use. Please contact your local bookseller or the Macmillan Corporate and
Premium Sales Department at 1-800-221-7945, extension 5442, or by email at
MacmillanSpecialMarkets@macmillan.com.

First Edition: 2025

10 9 8 7 6 5 4 3 2 1

*For Tracey, Abby, and Ellie, the sunshine in my life—
no matter the weather*

Contents

INTRODUCTION: *Doing Something About the Weather*   1

1. **TORNADOES**
   Widening the Window to Find Shelter   11
2. **FIRE**
   Watching the Wind, Stopping the Sparks   48
3. **THE LOCAL FORECAST**
   Inside Your Everyday Weather Report   73
4. **HYPERLOCAL WEATHER**
   The New Possibilities of Zooming In Tight   101
5. **EXTREME HEAT**
   How to Thwart a Silent Killer   128
6. **HURRICANES**
   A Planet-Wide View to Track Deadly Storms   164
7. **SEASONAL FORECASTING**
   Early Warnings for Droughts, Floods, and Famine   210

CONCLUSION: *Doing Even More About the Weather*   229

    ACKNOWLEDGMENTS   239
    GLOSSARY   243
    NOTES   245
    INDEX   265

CLOUD WARRIORS

Introduction

DOING SOMETHING ABOUT THE WEATHER

"Everybody talks about the weather, but nobody does anything about it." You may have heard that one before. The quote is often attributed to Mark Twain, though researchers have found evidence suggesting that some form of the aphorism originated with Charles Dudley Warner, Twain's friend and coauthor of their 1873 novel *The Gilded Age: A Tale of Today*.[1] As a quip, it resonates over the years because it seems to perfectly sum up our relationship with this omnipresent force of nature. Weather affects us all and we are constantly aware of it, but we cannot control or alter it. Nobody does anything about it, the saying suggests, because there isn't anything anyone *can* do about it.

Of course, that's simply not true. We can do a lot. Even if we can't control our planet's turbulent atmosphere, we can defend ourselves from its inconveniences and dangers. We can grab umbrellas on the way out the door when rain is likely or move an outdoor event inside in case of lightning. We can get people to shelter before a tornado sweeps through homes and schools. When a hurricane spins its way toward us, we can lay in supplies or evacuate while officials put utility crews on standby to repair downed power lines. We may not be able to stop the weather, but, contrary to the nineteenth-century witticism, we can do plenty to respond to its assaults. And now more than ever, we need to.

The world is on fire. As I write these words, people everywhere reel from a barrage of deadly and destructive weather events. Barely does one subside before another takes its place in the headlines. The past few years have run the gamut from too wet to too dry and everything in between.

An outbreak of tornadoes killed at least nine people in Alabama and Georgia—in winter. An ice storm left hundreds of thousands in Texas without power. Wildfires in Canada sent thousands fleeing from their homes and pumped smoke over a swath of North America, turning New York City into an eerie orange Martian landscape and shutting down airports due to poor visibility. Heat waves brought temperatures so high that, as of this writing, the World Meteorological Organization had pronounced 2023 the hottest year ever recorded. Fires ignited in Hawaii, razing a historic town in Maui and taking the lives of about a hundred people. The story repeats itself everywhere. A tropical cyclone brought catastrophic rains to Mozambique and Malawi, claiming more than fourteen hundred lives, then a heat wave lifted temperatures in India as high as 109 degrees Fahrenheit, resulting in dozens of casualties.

Though dangerous storms have always been with us, evidence now shows clear links between climate change and some types of extreme weather. In the case of wildfires, for example, climate change drives warmer and drier winds that suck moisture from vegetation, turning it into combustible tinder. A 2018 California state assessment warned that large fires could be expected to occur 50 percent more frequently by the end of the twenty-first century if high-emission climate scenarios came to pass.[2] Hurricanes, meanwhile, show evidence of gaining strength from warmer ocean waters; researchers say a greater proportion of tropical storms will reach category 4 or 5 hurricane status.[3] The need to address climate change is critical and urgent. But it's also important to recognize that many dangerous weather events aren't linked to the warming climate. Other factors increase the threats, such as building more homes in flood- and wildfire-prone regions. In other words, the repercussions of climate change provide additional compelling reasons to improve forecasting, but weather prediction offers a vital service regardless.

This is not a book about climate change; at least, not directly. This is a story about the weather—and the power of being able to see it coming. I don't want that focus to minimize the gravity of reducing emissions and combating the greenhouse effect. Quite the opposite; understanding extreme weather should make everyone more concerned about climate solutions.[4] But meaningful steps on slowing and reversing the growth of carbon dioxide in the atmosphere will require years to implement and decades before they can move the needle. We can't wait that long. Dangerous weather is here now, in many ways getting worse all the time. We need to elevate our ability to anticipate, prepare for, and respond to the weather we currently have. In 2023, the National Oceanic and Atmospheric Administration tallied a record twenty-eight separate weather disasters in which costs exceeded one billion dollars. With havoc like this bearing down, we must do better—and we can. In a war, you need intelligence to win. In our battle with the atmosphere, a good forecast provides that intelligence. It turns out that we can do something right now to save lives, reduce inequality, and safeguard the economy: Predict the weather better.

For centuries, humans have sought to understand and foretell the weather. Ancient Chinese chronicles show that observers kept detailed records of the weather in an effort to understand the vicissitudes of rain and wind. In the fourth century BC, Aristotle attempted to explain clouds and storms with the four supposed fundamental elements of earth, water, air, and fire. In the nineteenth century, modern forecasting entered what might be considered its first wave with the advent of the telegraph. Because weather patterns move predominantly from west to east, observers equipped with barometers and thermometers could message ahead to those in the east about conditions on the way. In the 1950s and 1960s, the Space Age helped usher in a second wave that constituted a mammoth leap. Satellites and radar provided unprecedented observations, while the availability of digital computers made it possible to plug all that data into complex equations that predicted the atmosphere's future behavior. Those keystones—satellite observations, radar data, and computing power—helped meteorologists achieve steady progress in forecasting in the subsequent decades.

Now, with remarkably little fanfare, a third wave of forecasting advances is arriving, driven by a number of new developments. Meteorologists have begun to tap artificial intelligence and machine learning to predict the weather. Oceangoing drones and new types of satellites provide valuable data to hone forecast models. The Internet of Things promises to turn everything from cell phones to cars into ubiquitous weather sensors. Equally significant in this third wave are new efforts to understand how people respond to forecasts and warnings. In the wake of some well-predicted yet nonetheless deadly events in recent years, meteorologists have been left to ask: How can good forecasts still see such tragic outcomes? To get answers, social scientists are studying how people get their weather information and how they use it, even visiting post-storm disaster scenes to ask survivors about their experiences and decisions. These researchers are learning about inequities that play a role in the lives lost in a storm. Language barriers and health issues can put some vulnerable groups at extra risk, while economic circumstances create obstacles to evacuations. Officials are using this research to improve communication to diverse citizen populations and prompt people to take the best steps for safety.

Weather predictions are already impressively good, so much so that their accuracy may surprise you. Consider the challenge: Anticipating the future state of Earth's atmosphere, a planet-wrapping blanket of gases that extends more than fifty miles above the surface before it begins to dissipate into space. Those gases move continuously at every moment in swirls of pressures and temperatures, constantly shaped by heat from the sun and the Earth and by moisture from the oceans. Imagine trying to predict the changing blooms of the creamer in your coffee as it mixes in. Venturing to foretell how such a complex system will change seems staggering, but—despite jokes and stereotypes of often-wrong TV weather presenters—current forecasting abilities provide substantial value.

Today's five-day weather forecast demonstrates accuracy at least as good as a one-day forecast from 1980, researchers have concluded. In other words, compared with forty years ago, we can look five times further into the future.[5] Progress looks equally striking when it comes to predicting some of the most dangerous threats. Take hurricanes, for example. When forecasters at the US National Hurricane Center (NHC) project a storm's future movement on a map—known as the track—seventy-two hours out, the average error these days amounts to less than a hundred nautical miles.[6] Those NHC forecasts have essentially doubled in accuracy compared with the forecasts of only twenty years ago.

These aren't just statistics. I remember the moment the significance crystallized for me. I was attending a discussion at the American Meteorological Society about the just-completed 2017 hurricane season. It had been a rough year due to three major storms: Harvey, Irma, and Maria. The assembled weather experts soberly reviewed the lives lost and the destruction, particularly in Puerto Rico, which would need years to recover from Maria. But they went on to celebrate a milestone in doing something about the weather. When the NHC forecast projected that Irma would threaten Florida, Governor Rick Scott declared a state of emergency when the fledgling storm was still a thousand miles and six days away from the Sunshine State. Part of the accomplishment was the accuracy of the forecast. But in a sense, the real achievement was earning the credibility that encouraged officials to act on such forecasts far in advance.

Predictions for other types of weather have demonstrated similar gains in prescience, from temperature outlooks that warn of dangerous heat waves to advance notice on the likelihood of thunderstorms capable of spawning deadly tornadoes. Each step forward creates opportunities for us to make smarter decisions and take action about the weather.

This book tells the story of the people bringing weather forecasting to the next level—and using those forecasts for the benefit of individuals, the economy, and society. In the chapters ahead, I will take you inside the progress underway in predicting some of the most hazardous and destructive weather, threat by threat. You'll see how each danger presents different problems. Tornadoes form swiftly in certain kinds of thunderstorms, requiring cutting-edge short-term forecasts to provide extra minutes of warning that widen the window for people to take shelter. Hurricanes, by contrast, take days to evolve from small atmospheric disturbances to fearsome storms that can rip the roofs from homes and flood entire cities. Forecasters must determine which of those disturbances will actually grow into tropical storms, then project the path of a storm and its strength days in advance to guide preparations and possible evacuations. And just as meteorologists consider different timescales, they must also zoom in and out by location. Health-jeopardizing heat waves can threaten people across hundreds of miles. But farmers with land situated in microclimates need hyperlocal forecasts so they can tell when rain will soak one field while leaving dry the crops growing on the other side of a hill.

Better forecasting can't provide all its potential benefits unless those affected by the weather know how to interpret and act on predictions. In conversations and visits with hundreds of scientists and forecasters,

I've come to believe that one of the most important skills to nurture is what some refer to as weather literacy.[7] Tracking forecasts and knowing the lingo is no longer just for professionals and weather nerds. Everyone needs to become smarter about the weather and savvier about cutting-edge forecasts. I hope this book can help.

Weather literacy looks different for different groups. For individuals, it can be as simple as learning more about the nature of each threat and the steps needed to stay safe. You may have seen images of a flash flood on TV, but have you truly considered how you would respond in those fast-rising waters? Weather literacy means taking notice of forecasts and having a plan. It may be obvious that it's good to seek refuge when the sirens of a tornado warning sound, but it's even better to be aware of the likelihood of a twister a day or two in advance and come up with a shelter plan. Weather communicators can help. For instance, meteorologists at the Weather Channel developed stunning computer graphics that displayed simulations of floods and storms superimposed over video of actual cityscapes, giving viewers a visceral, virtual-reality sense of the dangers. Of course, not everyone has access to this kind of information. Researchers increasingly recognize the impact of income, class, ethnicity, and culture on how people react to forecasts and warnings. Weather inequality is real, and it puts specific groups at greater risk.

Businesses and workers must improve their weather literacy too. Already, many companies give more attention to forecasts and outlooks than is generally recognized. Manufacturers pay to receive targeted alerts regarding key factories and facilities, while retailers look to seasonal forecasts to guide decisions on product orders from snow shovels to beach umbrellas. But the pressure grows, particularly in the context of extreme weather. Managers must take responsibility for outdoor workers in dangerous heat or cold. Emerging industries, from delivery drones to self-driving cars, must draw on pinpoint forecasts for safety and efficiency. Weather-focused companies ranging from big established players like AccuWeather to a host of lesser-known start-ups cater to all these needs.

Then there are institutions and government agencies that must shoulder the burden to improve their own weather literacy and promote a deeper understanding in their organizations and communities. Complicated weather-related decisions have to be made all the time by officials at every level, from small-town managers weighing road closures to school superintendents deciding whether to cancel classes to state governors considering evacuations and all the way up to advisers counseling the president about disaster declarations. As recent extreme events have demonstrated, every region must be prepared for practically any kind of weather, despite what locals have come to expect historically. The typically temperate Pacific Northwest has been overwhelmed by new and intense heat waves, while Texans have watched severe winter storms cripple their electric grid. In August 2023, Southern California found itself under its first-ever tropical storm warning prior to the arrival of Hilary, which flooded highways and shut down national parks.

Whoever you are—member of the public, manager in a business, leader in an organization, or official with a government agency—weather-literacy insights can help. This book will take you deeper into what you need to know, but here are a few broad principles. First, pay close attention to probabilities included in weather forecasts. Too often, people don't note the probability attached to a forecast and later conclude that it wasn't accurate. A 30 percent chance of rain means a 70 percent chance that it won't rain—but it's still likely enough that you should prepare for wet weather. Second, think about weather risks across multiple timescales and take appropriate steps at each stage. If you're in the projected path of a hurricane that's still days away, use the early notice to ensure you'll have enough food and batteries or even to make evacuation plans, then adjust efforts as the forecast firms up. Finally, know that, particularly in a changing climate, no one is immune from any type of weather. Even if you live where tornadoes are the best-known danger, make sure you know how to keep safe in flash floods and wind-driven wildfires and anything else that might materialize. In short, make the most of the forecast.

The efforts to build better forecasts span a fascinating mixture of science, technology, and fieldwork. To research this book, I went inside key organizations and companies in the world of meteorology. My reporting took me from storm-chasing with NOAA scientists in Oklahoma to a Colorado clean room to watch a new weather satellite getting readied for space. I met with veteran forecasters and spoke with students studying to be part of the next generation of meteorologists. I explored key companies—both big familiar names like the Weather Channel and small, innovative start-ups—shaping the future of forecasting. And I talked with the people who increasingly rely on weather intelligence, a surprisingly broad spectrum of users ranging from drone operators delivering packages from your local Walmart to individual farmers in West Africa struggling to coax more corn out of their fields.

Along the way, I encountered constant reminders of the challenges and the opportunities in the years ahead. Forecasting keeps improving, thanks to a cadre of driven scientists, programmers, and engineers who tackle the incredibly complex problem of understanding the Earth's atmosphere but toil largely in obscurity compared with, say, the innovators behind electric cars or cryptocurrencies. You'll find these people distributed among government agencies, research universities, and companies, a triad usually described in meteorology circles as the "weather enterprise."

The three legs of that tripod have propelled startling progress over the past few decades, though not without tension. Profit-seeking businesses have at times pushed to limit the government's role, essentially complaining that the government gives away too much (taxpayer-funded!) information. The push-pull between public and private forecasting faces new challenges as start-ups look to disrupt the field. Policymakers must nurture innovation in the private sector while supporting government work essential to public safety. Another hurdle may be tougher to surmount in these divisive times: the worsening politicization of science. Climate researchers have long suffered attacks from deniers and those whose interests are threatened by steps to lower carbon emissions. Increasingly, forecasters are also swept up in these onslaughts. In 2019, President Donald Trump roiled the weather world with an incorrect hurricane forecast map in what became known as Sharpiegate. Prior to the 2024 election, conservatives with ties to Trump called for breaking up NOAA and drastically upending the work of the National Weather Service.[8] It remains to be seen if such plans will gain traction in the second Trump administration. Given how much local officials rely on the Weather Service for public safety—work you'll read about in this book—legislators may be reluctant to sign on for big disruptions regardless of their political stripes. The US weather enterprise has produced extraordinary progress and heralds more improvements to come—but only if science leads the way.

As I probed further into the future of forecasting, the people of the weather enterprise helped draw me more deeply into this story. Of course, you cannot paint an entire group with a broad brush. Some are motivated by pure science, while others are drawn by entrepreneurship and opportunities to profit. Yet so many of the people I came to know spoke earnestly and unironically of advancing goals that are perhaps best summed up by two of the National Weather Service's guiding principles: saving lives and protecting property. After a while, I recognized some shared values that resonated with my own profession. Though journalism manifests undeniable flaws and debatable choices, so many of the people I've been fortunate enough to work with in my career care keenly about benefiting society by reporting and telling stories that demand attention. I perceived a similar sense of mission across the diverse people who play roles in weather forecasting. Though it may not be entirely prudent to invoke a martial metaphor, I came to think of them as warriors—people with a concept of duty and an obligation to protect—fighting a battle against the elements. Their opponent, one that can be guarded against but never ultimately defeated, is the pervasive and ethereal power of the atmosphere itself—hence my fanciful conception of them as cloud warriors. But no matter what you call them, these legions of mostly unseen guardians defend humanity against dangerous weather.

Above all in my research and travels, I gained a deeper appreciation of the urgency of turning better forecasts into smarter actions—into *doing* something about the weather. I was surprised to learn that much of the work on closing that gap is relatively recent. For a long time, meteorologists focused on producing the best and most accurate forecasts possible. But as predictions got better, it raised a difficult question: If weather forecasts are so much better, why are so many people killed and injured in some events? The shift to studying people as well as the atmosphere resulted in such pioneering work as the 2015 project VORTEX-SE, in which US meteorologists and social scientists worked together to examine tornado tragedies in the Southeast. Researchers discovered that no matter how good the warnings were, some individuals weren't aware of them or didn't have access to appropriate shelter—a problem exacerbated by income disparities and other social issues. Insights from this kind of work have rippled throughout the meteorology world. When I visited local National Weather Service forecast offices, I saw firsthand how meteorologists are working to address these concerns, from constantly communicating with first responders and hospitals to directly reaching out to vulnerable communities. These efforts underscore the connection between seemingly esoteric science and people's day-to-day lives—and serve as a force multiplier for better forecasting.

In 2023, an art exhibition opened in Venice in the historic Ca' Corner della Regina palazzo on the Grand Canal. It was titled *Everybody Talks About the Weather*. The Fondazione Prada, which sponsored the show, described it as a chance to "trace the various ways in which climate and weather have shaped our histories and how humanity has dealt with our everyday exposure to meteorological events."[9] At the entrance, a giant screen showed television weather forecasters with their maps. Inside, visitors encountered a mix of old masters' depictions of storms and snow with contemporary multimedia works, such as a sculpture formed from lightning-rod balls.

The thrust of the exhibition wasn't hard to understand. Living in a time of climate change, people *should* talk about the weather. And, further, action is needed: We must do something. In a sign of our times, the message of *Everybody Talks About the Weather* turns the nineteenth-century aphorism on its head. Instead of shrugging at the futility of changing the weather, we must embrace our agency and respond to it. The state of the climate itself, of course, demands initiative; lowering carbon emissions is a crucial and decades-long project requiring everything from better-informed individual choices to arduous global diplomacy. Against that backdrop, though, weather dangers will keep on coming. Anyone can wind up in the crosshairs. More accurate forecasts provide the basis to act. Improving weather literacy can rebut the notion that we're helpless. We can make plans for what to do when dry winds raise the threat of wildfires or when a tornado requires us to find immediate shelter. We can get ready for extreme heat waves and prepare our go-bags in case floodwaters threaten.

All of this means paying more attention to the weather, taking in the predictions, and reaping the benefits of the immense progress happening now in meteorology. Keeping people safe. Minimizing the damage and costs. Enabling new industries. It all starts with a better forecast.

1 Tornadoes

WIDENING THE WINDOW TO FIND SHELTER

As our pickup truck careens north through Oklahoma on US 283, storm clouds descend in a blue-gray curtain. Here in the Great Plains, the relentlessly flat terrain can make the sky feel endless. But not on this early May afternoon. Less than a mile ahead, the two-lane highway disappears into darkness. Even more ominous than what we see is what we hear. Inside the truck, a rhythmic thumping envelops us. The sky is falling, one icy chunk at a time, as the storm drops a torrent of hail. Staccato pounding turns the pickup's crew cab into the inside of a snare drum.

This is tornado season in the heart of Tornado Alley—and our truck is barreling right into the storm. The driver, a scientist named Sean Waugh, treats the alarming conditions as practically routine. In a sense, for him, it is. Waugh is one of the most experienced storm-chasers at the National Severe Storms Laboratory (NSSL) in Norman, Oklahoma. When forecasts call for threatening weather, it's Waugh's job to plunge into the thick of it. The goal: collect measurements to help meteorologists better understand tornadoes and, hopefully, give people earlier warnings to get out of harm's way.

Waugh started out storm-chasing as an undergrad back in 2005. Outfitted in a black T-shirt and khaki cargo shorts and sporting a wispy beard, Waugh looks more like a relaxed college junior than an accomplished Ph.D. Throughout our expedition, he jokes and laughs and occasionally obsesses about where to find lunch. He considers coming across a Taco Bell to be a lucky omen for intercepting an interesting storm. But when he talks about the puzzle at the heart of his work, Waugh turns intense. Every bit of data gathered contributes to our understanding of destructive storms, he says. "What causes this storm to produce a tornado? Another way of thinking about that is: What causes this storm to not produce a tornado?"

The charcoal-gray pickup we're riding in—dubbed Probe 2 by the NSSL team—began its life as a run-of-the-mill 2014 Ram 1500. But Waugh and his colleagues have transformed it into a mobile weather research station. For starters, Probe 2 is hail-resistant—a good thing, given the pummeling we're taking. The hail shield is low tech: a cage-style heavy wire screen. It starts at the back of the vehicle, where a cargo cap encloses the truck bed, covers the roof of the passenger cab, and stretches forward over the driver's head to protect the windshield and the full length of the engine compartment; it's supported by posts bolted to the hood and fenders. The scientists have mounted an array of instruments on top of this structure, including devices to measure wind speed, direction, temperature, and humidity. The contraption looks like someone went crazy with a kid's Erector Set and some PVC tubing; the spindly instrument stack cantilevered out over the hood effectively doubles the height of the truck. We draw stares from drivers of other cars as we pass by.

Inside the truck is the nerve center for all this instrumentation. A laptop computer, mounted between the driver and front passenger, displays and records the constant stream of data from the atmospheric instruments. The laptop also tracks the truck's speed and direction, adjusting the raw data from the wind sensors to subtract the effect of the vehicle's movement. Riding shotgun with Waugh today is Kiel Ortega, a research colleague who keeps an eye on the measurements pouring into the laptop. A GoPro camera on the dashboard records the view ahead.

Waugh and Ortega periodically stop their race across Oklahoma in pursuit of developing storms to launch weather balloons, five-foot-wide orbs they inflate from a rack of helium tanks mounted in the back of the truck. The pickup's Rube Goldberg weather station can record observations only at ground level. Each helium balloon dangles a small package of instruments called a radiosonde that samples the atmosphere continuously as it rises and transmits data at every altitude, effectively taking a core sample of the sky. Wedged next to me in the pickup's back seat, a second laptop receives and graphs the radiosonde data. The balloons can soar up to 100,000 feet. Eventually the balloon will burst in the thin air. At that point, the lightweight instrument package, a bit smaller than a one-quart milk carton, drifts back to Earth. If you happen to find one, you'll discover attached instructions for how to mail the gadget back.

Waugh and his colleagues call their truck a mobile mesonet. In meteorology, the word *mesoscale* describes events that occur over distances from a few to several hundred kilometers in the atmosphere, such as thunderstorms and so-called dry lines that separate masses of moist and dry air and help storms get started.[10] Hence the term mesoscale network, or mesonet for short. Weather researchers have created ground-based stationary mesonets all over the country. Oklahoma pioneered mesonets with a network of 120 automated stations. There's at least one in every county in the state, so as we drive around Oklahoma, we're never too far from one. But scientists at NSSL need an even closer look, and the idea of the mobile mesonet is to put sensors right where the action is. Waugh and his colleagues have taken Probe 2, and its sister truck, Probe 1, over wide stretches of the country. When a forecast calls for consequential weather, they hit the road, sometimes for days or weeks.

As hail hammers Probe 2 with an unnerving crescendo, Waugh steers us toward a thunderstorm throwing down what's known as a wall cloud. It's a distinctive and dramatic formation that can augur a tornado. The dark shroud looks cataclysmal, particularly considering we were driving under sunny skies just an hour earlier. Out the window, big chunks of ice hit the road and bounce back up like popcorn popping. Inside, the front-seat laptop displays data streaming in from the mobile lab's sensors. Will a tornado materialize from this storm? Can we see it coming? "The only way to figure that out is to keep making observations," Waugh says. That's why we're out here today.

A tornado is one of nature's most frighteningly capricious threats. It can form in minutes and dissipate just as quickly—but it can also persist for an hour or longer. With winds anywhere from forty miles an hour all the way up to two hundred miles an hour or more, a tornado can suck a tractor-trailer into the air, derail a train, and lift a house off its foundations. The most intense tornado winds recorded to date were observed in a 1999 twister that ravaged the Oklahoma City area. A mobile Doppler radar unit measured winds over three hundred miles per hour in that storm.[11] Tornadoes can etch a path of destruction that snakes across a landscape, shredding some homes and buildings while leaving others freakishly untouched. They can kill dozens or sometimes hundreds in a single strike, or they can appear in swarms, dipping down and up and dotting an area with destruction.

The deadliest recorded tornado in US history, known as the Tri-State Tornado, occurred in 1925 and claimed an estimated 695 lives as it ripped across southeast Missouri, southern Illinois, and the southwest tip of Indiana. More recently, the 2011 Joplin, Missouri, tornado killed 158 as it carved its way across twenty-two miles. Damage from the Joplin tornado was assessed at $2.8 billion.[12] Some years, the deaths from tornadoes are measured in dozens; other years, hundreds are killed. The difference between safety and tragedy can depend on everything from the intensity of the storm and the durability of the local buildings to how quickly people learn that a tornado is on the way and whether they reach shelter in time.

Meteorologists are very good at detecting tornadoes. They're also fairly good at predicting when the weather will be favorable for tornado formation, even a few days out. But between those timescales, there remains much to learn. Forecasting a tornado represents one of the most formidable challenges in meteorology. Consider this simple fact: Tornado warnings are not forecasts—they are observations. Depending on where you live and how often you need to watch out for tornadoes, you may or may not be aware of this. But when the National Weather Service (NWS) issues an official tornado warning, meteorologists aren't making a prediction—they have either detected the telltale rotation of air on their radar screens or someone has actually sighted and reported a twister.

The upshot is that, according to NWS statistics for recent years, the average warning lead time for a tornado in the United States is about nine minutes.[13] That means nine precious minutes for people to take cover in their basements and nine minutes for people outdoors to make their way to a sturdy building—assuming they've actually heard the warning. If you're away from home, it means nine minutes to locate and get yourself to a safe shelter. And though false alarms for tornado warnings have decreased, they still hover just below 70 percent (meaning that for about 70 percent of warnings issued, no tornado is subsequently verified).[14] That number might surprise you with how high it is. And you might understandably wonder how there can be any false alarms at all, given the observation-based nature of tornado warnings. But radar indications of rotation in the air don't always result in a tornado that touches the ground; spotters can also be mistaken in their reports.

It's not hard to see that nine minutes can be a slim margin depending on where you happen to be, what you're doing, and whether you learn about a warning as soon as it's issued. If you're a suburbanite with a job that's not in your home, imagine you are somewhere on your commute back from work when you hear a tornado warning on the radio or receive an alert on your phone. A car is considered a dangerous place to be in a tornado. The intense winds can fling heavy debris at the vehicle and potentially even lift the car itself. With the clock ticking, can you think of a suitable building within a few minutes' drive, proceed there calmly, and take cover in time? In suburban areas, it's most likely doable. But if you perform the thought experiment, you may realize that considering in advance what you would do and where you would go might be a big help.

In the case of large and deadly twisters, warning times are likely to be longer and provide you with an additional few minutes. National Weather Service statistics show that if the smallest tornadoes—which account for relatively few fatalities—are excluded, average warning times rise to about fourteen and a half minutes.[15] In any case, doubling warning times would represent a leap forward. The National Oceanic and Atmospheric Administration (NOAA), the parent agency of the NWS, is under a congressional mandate to do just that—and more. In 2017, Congress passed the Weather Research and Forecasting Innovation Act, which President Donald Trump signed into law on April 18 of that year. It requires the Commerce Department, the cabinet-level agency of which NOAA is a part, to "establish a tornado warning improvement and extension program."[16] The goal, as the law states, is to "reduce the loss of life and economic losses from tornadoes through the development and extension of accurate, effective, and timely tornado forecasts, predictions, and warnings, including the prediction of tornadoes beyond 1 hour in advance."

That's a tall order. Science doesn't work on a timetable. But progress is ongoing.

The work to protect people from tornadoes takes place on two major fronts. The first strives to add minutes to the warning time. That means understanding the conditions that spawn a tornado—or, as Waugh put it on our storm-chase, figuring out why one severe thunderstorm produces a tornado while another doesn't. On this front are physicists, atmospheric scientists, engineers, and more. It calls for a sizable paradigm shift, from the current warn-on-detection method to a warn-on-forecast approach. To make this shift, meteorologists will need to not only predict the emergence of a tornado before it forms but also have enough confidence in those predictions to feel comfortable alerting the public.

The second front is more akin to that thought experiment about what you would do in the event of a tornado. This work focuses on understanding the barriers that keep people from taking the best steps when they're in danger from a tornado. The leaders on this front include experts in the social sciences, such as psychologists, economists, and sociologists. Advancing this work means learning the different needs of different groups, which depend on where they live, what language they speak, what resources they have, and how they perceive the risk.

Figuring out when a storm is coming and getting the right information to the right people so they can do the right thing—it all needs to come together to protect people from tornadoes.

A tornado is defined almost universally by meteorologists as "a rapidly rotating column of air that touches the ground." You might call that a textbook definition, as that is, in fact, how it's defined in *Meteorology Today,* a popular college introductory volume written by C. Donald Ahrens and Robert Henson.[17] To understand a tornado's formation, let's back up a bit and consider some thunderstorm basics. A thunderstorm is a phenomenon of convection—the movement of heat and moisture. We know that warm air rises because it's less dense.[18] In the atmosphere, when warm, moist air rises, it can form a thunderstorm. The rising warm air creates an updraft. As the air rises, it cools, and the moisture condenses. Eventually the heavier air sinks, creating a downdraft. Put the updraft and the downdraft together and you've got a thunderstorm. That's a simplification, but it underscores the fundamentals. All that moving air creates turbulence; conditions are ripe for rain or hail.

For tornado weather, we're concerned mainly with a specific type of thunderstorm called a supercell. If the winds are moving in one direction nearer the ground and a different direction above it, they may form a rotating tube of air parallel with the ground. If the updraft knocks that tube into a tilt, the result can be a large vertical area of rotating air. When a big thunderstorm has such a mass of rotating air—technically known as a mesocyclone—that's a supercell. Sometimes the falling cold air from above will form downdrafts toward both the front and the back of the storm as it moves and create a configuration that causes the rotating air in the heart of it to spin faster and faster, narrowing and touching the ground: a tornado. Though supercells, with their big rotating mesocyclones, are responsible for most tornadoes, other storm configurations can generate twisters. Squall lines—storms strung out in a relatively thin band but stretching as much as hundreds of miles—can produce a smaller area of rotation called a mesovortex. These mesovortices can also result in tornadoes. Meteorologists often refer to squall-line storms and related formations as a quasi-linear convective system, or QLCS. Forecasting tornado formation in a QLCS is considered more difficult than forecasting it in a supercell.

Starting in 1971, tornadoes were classified by the Fujita scale, a framework named after its creator, weather scientist Tetsuya Theodore Fujita. Fujita, who died in 1998 after a long career at the University of Chicago, had a nickname that speaks volumes about his influence in meteorology: "Mr. Tornado." Fujita was a true pioneer, and the path that led to his important contributions is well known among weather buffs. Fujita was born in Japan in 1920, studied at what was then the Meiji College of Technology (now the Kyushu Institute of Technology), then joined the faculty there. In *Memoirs of an Effort to Unlock the Mystery of Severe Storms,* his autobiography, Fujita recounted a formative early moment in his work.[19] Shortly after the end of World War II, the president of Fujita's institution dispatched him, along with other faculty and student volunteers, on a "ground-truth mission" to survey the two cities on which the United States had dropped atomic bombs. Fujita describes his efforts to figure out where the bomb detonated over Nagasaki. He recalls visiting a cemetery and noticing a flowerpot that displayed a "crescent-shaped burn mark on the bamboo surface" and realizing it could help estimate the position of the detonation. "I hopped around from one cemetery to another on the hillsides north, east and south of ground zero," Fujita wrote in 1994. "With a high degree of accuracy, I estimated the fireball height to be 520-m AGL [meters above ground level]." He added, "I am glad that I have not been adversely affected by the remnant radiation to which I had been exposed during those two research visits in September 1945." The work on the bomb damage proved useful to him in understanding the effects of thunderstorm downbursts, when a powerful wind from above pushes down toward the ground.

Some years later, Fujita was, as he put it, "fished out of postwar Japan"; he moved to the United States to work with the University of Chicago's Horace Byers, head of the school's meteorology department. Byers himself is a formidable figure in the history of weather forecasting, having led a program mandated by Congress in 1945 to better understand severe weather, particularly in the context of safety for the burgeoning aviation industry, called the Thunderstorm Project.[20] Fujita referred to Byers as "my fatherly mentor professor." At the University of Chicago, Fujita went on to do groundbreaking work on severe-weather phenomena, including hurricanes, thunderstorms, downbursts, and—of course—tornadoes. His development of the Fujita scale stemmed from research that hearkened back to his surveys of Nagasaki and Hiroshima. Fascinated by the level of destruction wrought by tornadoes, he sought to analyze the damage. After an outbreak of Midwest twisters in April 1965 known as the Palm Sunday tornadoes, Fujita conducted aerial surveys—riding in a Cessna to view and photograph fields ripped apart by the tornadoes—and visited sites on the ground. Fujita wanted a way to classify tornadoes by the level of damage they inflicted. Eventually he developed a system of categorizing individual tornadoes. The Fujita scale, introduced in 1971, rated tornadoes on a scale of 0 to 5 based on the observed damage and inferred wind speeds. A Fujita scale 0 tornado, or an F0, was designated as a twister with little damage and winds from 40 to 72 miles per hour. At the far end of the scale, Fujita placed the F5 tornado; with winds from 261 to 318 miles per hour, it was able to sweep away frame homes and rip crops out of the ground. But despite the precision of the wind-speed ranges, they are essentially guesses; the scale is rooted in classifying the damage from a tornado rather than directly measuring the winds.

Since 2007, US meteorologists have used a successor to Fujita's original system, the enhanced Fujita scale. The EF scale refined the damage assessments used to classify twisters and revised the associated wind estimates to reflect experience that showed lower wind speeds were capable of inflicting greater destruction than indicated by the original Fujita scale. Damage is assessed according to a point system using twenty-eight detailed indicators, such as a mangled service-station canopy or the destruction of a motel.[21] In keeping with the earlier system, tornadoes are classified from EF0 (65 to 85 miles per hour in a three-second gust, rather than sustained winds at that level) up to a monster EF5 (gusts over 200 miles per hour). But the fundamental approach—a scale that allows classification of tornadoes—remains the one formulated by Ted Fujita.

The United States is ground zero for tornado research like Fujita's because the country is essentially ground zero for tornadoes. The United States experiences more tornadoes than any other country. Though tornadoes occur in many places around the world, the geography of the United States is particularly favorable for them. Over the broad flat expanses of the Great Plains, cold and dry air from Canada at higher altitudes collides with warm, humid air moving north from the Gulf of Mexico. Those collisions create the type of instability that fuels thunderstorm and supercell development. Because northward-moving warm air plays a key role, spring and summer see the most tornadoes, though they are possible year-round. According to NOAA, the United States averages more than a thousand tornadoes each year. Also tornado-prone— though far less so—are Canada, parts of Europe, and Argentina. Middle latitudes, between 30 degrees and 50 degrees (in both the Northern and Southern Hemispheres) are prime territory because of the convergence of colder polar air and warmer, wetter equatorial air. And of course, much of the continental United States falls smack in that range. One bit of tornado trivia: On a tornadoes-per-square-mile basis, the United Kingdom ranks first, though most UK twisters—mainly seen in England—are relatively weak.[22]

But for sheer number of storms, the United States can't be beat. The portion of the Great Plains from north Texas up through Oklahoma, Kansas, and Nebraska sees so many twisters that it has earned the nickname Tornado Alley. There's no official designation of its boundaries; some would include additional states. It would be a mistake to believe the risk is confined to that zone of the country, however. The southeastern states, particularly Louisiana, Arkansas, Mississippi, Alabama, Georgia, and Tennessee, also experience a disproportionate number of tornadoes. Historically, according to NSSL, Tornado Alley suffers the heaviest action in May and June. The Gulf Coast states see the threat ramp up earlier in the spring or even in late winter. They're also more prone to nighttime tornadoes, which present some additional risks; people might not receive warning information because they're asleep, or if they're outdoors, they might not see signs of a brewing storm in the darkness.[23] Despite these geographic generalizations, any severe-storm meteorologist would caution that no place is completely immune. Tornadoes in the Pacific Northwest and New England, for instance, are relatively rare, but they do happen. And people living in those areas might be less familiar with the need to respond to warnings, putting them at higher risk when a tornado does make an appearance.

There's some evidence these geographic patterns are changing. In 2016, Purdue University scientist Ernest Agee and colleagues published an analysis of US tornadoes from 1954 to 2013 in the *Journal of Applied Meteorology and Climatology*. They found "a temporal shift of maximum activity away from the traditional Tornado Alley."[24] The center of activity, according to their findings, appeared to be moving east, with "increases concentrated in Tennessee/Alabama." They also described a relative slowdown in summer and an uptick in fall. Noting warmer-weather trends in more recent years, they suggested that climate change might be reshaping the map of tornadoes in the United States.

We already know quite a lot about tornadoes. We know the conditions that create thunderstorms, and we know that supercell storms are more likely than other storms to result in twisters. We know where tornadoes occur most frequently and why. Thanks to massive research efforts, we're learning more about them all the time. But we still don't know precisely why one storm spawns a tornado and another, very similar storm does not—the key question that Sean Waugh posed to me while piloting Probe

2 across western Oklahoma. This is the study of tornadogenesis—the process by which a tornado forms. The more we can understand, the better we can predict—and have enough confidence in those predictions to take actions based on them.

The epicenter of tornado research can be found in the city of Norman, Oklahoma. About twenty miles south of Oklahoma City, Norman is one of America's classic college towns, home to the University of Oklahoma, or OU, as it's known. Life in Norman means football Saturdays in the fall, restaurants and shops in the historic Campus Corner District, and sunny afternoons on the South Oval—all brimming with the energy of 29,000 undergrad and grad students. Norman has other, more relevant distinctions: Along with Oklahoma City proper, the college town falls within Tornado Alley. And OU operates one of the most respected meteorology programs in the world.

To find the hub of weather research, you need to wander a few miles south from the heart of campus. There you will come upon a sprawling building with office wings clad in brick and stone stretching out from a central glass atrium, all topped by an observation deck and a spire. This is the National Weather Center (NWC). The sign out front announces one of the facility's defining characteristics, a synergy between government and academic weather work. Flanking its name are two badges: the seal of the University of Oklahoma and the NOAA logo. This is a place for weather research and forecasting operations, academic endeavors and government commitments. The NWC opened in 2006 to provide a collaborative hub for all the meteorology work taking place in Norman across both the university and key NOAA divisions. The building itself, with 244,000 square feet of space, has five main floors, though its height is more in line with a nine-story building, thanks in part to space needed between the floors to contain a reported 3,200 miles of cables.[25]

Here is just a partial list of the groups and organizations working away at the National Weather Center. For starters, there is the University of Oklahoma's School of Meteorology and its parent group, the College of Atmospheric Sciences. Among the NOAA contingent, there is the NSSL, the premier government research center for tornadoes, severe thunderstorms, and related phenomena. There's also the Storm Prediction Center (SPC), where meteorologists provide forecasts for tornadoes and severe storms—including probabilities of hazardous weather days in advance—for the entire contiguous United States. Also located here is the National Weather Service's Norman weather forecast office (WFO), the source of everyday weather information for most of western Oklahoma and portions of north Texas. Then, spanning both OU and NOAA, there's the Cooperative Institute for Severe and High-Impact Weather Research and Operations (CIWRO), which facilitates joint university/government research. And in addition to all that are some state-focused weather operations, such as the Oklahoma Climatological Survey and the Oklahoma Mesonet, which oversees a network of local weather observation sensors.

Along with an abundance of brainpower and acronyms, the NWC facility offers some weather glitz. The atrium is dominated by a six-foot-diameter globe, called Science on a Sphere, that can display computerized projections of everything from satellite views of Earth to images of the surface of the sun. A nearby exhibit pays tribute to the 1996 film *Twister*, which starred Helen Hunt and Bill Paxton and introduced a generation of moviegoers to the concept of storm-chasing. In the movie, two teams of researchers race to be the first to place a sensor-packed cylinder in the path of an oncoming tornado. The NWC lobby displays the props used to represent those fictional probes—dubbed Dorothy and D.O.T. 3 in the film—as well as a third device: the nonfictional TOTO (an acronym for "*To*table *To*rnado Observatory"), a fifty-five-gallon orange drum with some weather sensors on top that helped inspire the plot of *Twister*. TOTO was a real-life instrument that NSSL and OU researchers used in the 1970s and 1980s to gather tornado measurements. Unlike Hunt's and Paxton's characters, researchers never got TOTO in perfect position, though they came close during an April 1984 tornado in Oklahoma. "It turned out that TOTO had a center of gravity which was too high for extreme wind, and fell down [. . .] as it was sideswiped by the edge of a weak tornado," according to NOAA.[26] The movie also included a memorable (but computer-generated) scene of a cow being swept through the air by a tornado, and today at the National Weather Center, students and researchers can grab burgers at the Flying Cow Café, an on-site cafeteria complete with a cartoony logo of a soaring bovine. (When Hollywood revisited tornadoes with *Twisters* in 2024, filmmakers called on NSSL meteorologists for weather expertise. Storm-chaser Sean Waugh, who consulted on science and technology, can be spotted in one scene.)

Twister may have been NSSL's shining moment in popular culture, but its other legacies have had far greater impact on forecasting and public safety. At each of 156 locations across the United States and its territories, from Riverton, Wyoming, to Evansville, Indiana, to Philadelphia, Chicago, and Los Angeles, you can find a thirty-nine-foot-wide white dome perched atop a metal tower. These stations—known as WSR-88D (for "Weather Surveillance Radar, 1988 Doppler") or NEXRAD (for "Next Generation Weather Radar") units—provide critical views of weather and are able to detect winds, rain, snow, and hail. They trace their history to NSSL research on using radar to detect severe storms and tornadoes. The key was Doppler radar. Ordinary radar can detect objects in the sky by beaming out radio waves and analyzing reflections. When the waves hit an object, they bounce back toward the radar transmitter, revealing distance and bearing. Doppler radar takes advantage of the Doppler effect, in which a wave's frequency is compressed or stretched based on whether an object is moving toward or away from an observer.[27] Doppler radar uses this phenomenon to measure the motion and velocity of precipitation in the air, providing valuable data on a storm's movement and speed. NEXRAD is one of the most important tools in day-to-day weather forecasting and severe-weather detection, and it owes its existence to NSSL research. The FAA depends on NEXRAD for aviation weather too. It's worth noting that this next-generation radar began rolling out operationally in the 1990s. Though NOAA continues to upgrade the WSR-88D capabilities, many believe the United States should move to a new generation of technology, the phased-array radar. NSSL and OU research has demonstrated that phased-array beams, which can be focused electronically on specific areas, can provide better information on severe storms and tornadoes.

One other legacy bears mentioning when it comes to the National Weather Center, NSSL, OU, and Norman: the tragedies that the surrounding community has suffered from tornadoes. On May 3, 1999, the southern Oklahoma City area was struck by one of the most powerful twisters ever recorded, part of a major outbreak that day and the next. The big one was a devastating F5 that endured for nearly an hour and a half, starting near Chickasaw and ripping through the suburbs of Bridge Creek and Moore as well as parts of Oklahoma City. NOAA later concluded that the single F5 tornado was directly responsible for thirty-six deaths. Moore is almost directly between Oklahoma City and Norman, a short drive from where the National Weather Center now stands. Fourteen years after that twister, on May 20, 2013, Moore was hit again, this time by an EF5 that killed twenty-four and injured over two hundred. A dramatic image of that tornado appeared on the cover of that week's *Time* magazine. When I visited the National Weather Center, it was striking to see that cover—which I had worked on years earlier with my *Time* colleagues—displayed on a wall in the NOAA offices along with other images memorializing that day. It reminded me that the researchers in Norman didn't just study tornadoes. They lived with them and with the sometimes heartbreaking consequences for their community.

The work at the National Weather Center is unquestionably tied up with the history of Tornado Alley and those deadly storms. But when I visited, the spring tornado season was getting underway, and the focus was on the future.

"Do we believe this?" Michael Stroz asks while scanning the readouts on his computer display and considering whether to issue a tornado warning. It's 4:14 p.m. here in Norman and the "this" in question is an update from a colleague passing on a possible sighting of an incipient tornado. Three minutes earlier, a storm spotter had reported seeing a low cloud with a funnel shape near Gorham, Kansas.[28] Spotter information can be hugely valuable, but in this case, Stroz must judge whether the report jibes with the data on his screen. "It's not supported," he concludes.

Stroz is an experienced NWS meteorologist, accustomed to making such calls. But today, his job is different. Sitting at a workstation in NSSL's suite of offices on the second floor of the National Weather Center, Stroz is keeping watch over a slice of Kansas as part of an experiment. He sifts through a constantly updating display of radar and satellite data, eyes flickering over sixteen different windows open at once across two side-by-side computer monitors. Conditions are ripe for severe thunderstorms today. If Stroz sees evidence of a tornado spinning up, he will issue a tornado warning—but it will go nowhere beyond this room. If a tornado does appear, the people of Gorham, Kansas, will be alerted by meteorologists in their local weather forecast office, located in Wichita.

Stroz and a group of colleagues are stationed here in Norman today as the weather world's equivalent of test pilots, tasked with trying new forecasting tools to see if they merit rolling out to everyone at the Weather Service. Though their evaluations of the weather and decisions about warnings won't go out to the public, they treat the work just as seriously. It's a pressure test for potential improvements. These forecasters are participating in an annual ritual known as the Spring Experiment, part of what NOAA calls the Hazardous Weather Testbed. Each year at this time, selected Weather Service meteorologists take a break from their day jobs in forecast offices around the country and converge on Norman to collaborate at NSSL. They are joined by researchers, engineers, and others working to improve forecasting. Each spring brings a new list of topics to study, and this year is no exception.

Today Stroz is watching the lightning, thanks to the latest generation of weather satellites. On November 19, 2016, an Atlas V rocket blasted off from Cape Canaveral carrying the next-generation craft in the joint NOAA/NASA Geostationary Operational Environmental Satellites (GOES) program, designated GOES-R. Once it reached orbit, it became GOES-16.[29] A geostationary satellite operates in an orbit that keeps it in place relative to the surface below—very useful for a weather satellite. Once it was in orbit 22,236 miles above the Earth, GOES-16 underwent a series of tests, after which it was moved into a prime position to view the eastern half of the United States, replacing an old satellite and assuming the official mantle of GOES-East. The western side of the country is covered by GOES-West.

The satellite that GOES-16 replaced as the sentinel for the eastern United States had been launched in 2006, so the new bird contained years of technological improvements over its predecessor. The old satellite—I won't detail all the renamings it has endured—was moved and is now used by the US Space Force to monitor weather over the Indian Ocean. The most attention-getting instrument on GOES-16 is the Advanced Baseline Imager (ABI), a high-tech camera that captures pictures of Earth at a faster rate and with four times the resolution of the previous satellite. The ABI produces photos of our planet that are gorgeous, even awe-inspiring. It also sees things the human eye cannot—infrared and near-infrared bands that detect moisture and smoke and provide data critical for forecasters and models.

GOES-16 carries another new instrument. It has not drawn as much notice as the ABI and its lush Earth pictures, yet it offers intriguing new information to forecasters. The Geostationary Lightning Mapper (GLM) can *see* lightning even during daylight, thanks to an optical sensor tuned to a specific near-infrared band. Lightning, of course, represents a weather danger by itself: According to the NWS, the United States saw an average of twenty-seven reported deaths a year from lightning from 2009 to 2018. But knowing where the lightning is can also help forecasters understand what's happening inside a storm. One particular phenomenon known as the lightning jump, a sudden, sharp rise in lightning activity, often precedes severe thunderstorm hazards, including hail and tornadoes.

For scientist Hugh Christian, the launch of the GLM on GOES-16 represented the culmination of more than three decades of work. "It's been a long slog," he tells me. Christian is a physicist who started working for NASA in 1980 at the Marshall Space Flight Center in Huntsville, Alabama. He was hired to focus on lightning observation in space. "The whole question was how in the world could you detect lightning during daylight," Christian recalls. Doing so required not only seeing light at the appropriate wavelengths but also capturing images very quickly, given the brief span of each lightning pulse. Christian and his colleagues tested the concept with instruments on a U-2 aircraft. The U-2 is best known as a spy plane, but the same ability to fly at high altitudes—some 70,000 feet, or twice as high as a commercial jetliner—that made it useful for surveillance also made it an ideal research platform. The tests paid off. "We finally decided it was possible," he says. "Then came the problem of trying to convince people who have money to spend money on it. In those days, meteorologists didn't think lightning was of value in forecasting. We had to do the science to prove the value of it."

Over time, Christian and other researchers demonstrated that the sudden jumps in lightning corresponded to storms and rain. They found that the rate of lightning flashes was an indicator of the amount of energy within a storm cloud. "The more energy you have, the more severe the storm. We can get an idea of whether a storm is intensifying or decreasing," he says. "We were able to show that as a storm starts to form, there's a sharp increase in lightning-flash rate. We call that the lightning jump." In 2006, Christian moved from NASA to the University of Alabama in Huntsville, where he continued his research and worked with NASA, NOAA, and Lockheed Martin, the defense and aerospace giant that built GOES-16. The result was the GLM instrument. Christian was on hand at Cape Canaveral that night in November 2016 when the Atlas rocket lit up the sky and lofted GOES-16 into orbit. "It was incredible," he says. "I put thirty-seven years, basically my career, into this. It was a matter of doing science, of doing technology. And also of salesmanship." Christian believes lightning observations from space are a game-changer. "It's going to have an impact and save lives."

But there's a lot of work needed to turn the GLM's observations into a usable and reliable tool for NWS meteorologists. This work is often referred to at NOAA as "R2O," for "research to operations." That's where Stroz and the Spring Experiment come in. Joining him at NSSL are Samantha Edgington, a physicist from Lockheed Martin in charge of calibrating and assessing the GLM; her Lockheed colleague Clem Tillier; and Christopher Schultz, a lightning-jump expert and NASA scientist who wrote his Ph.D. dissertation on the subject. All are collaborating with Kristin Calhoun, an NSSL researcher overseeing the GLM work at the Hazardous Weather Testbed. The setup allows Stroz to ask questions of the GLM experts while trying to factor the lightning data into his assessments of the weather over Kansas. All can then try to devise improvements and guidance—even while the Spring Experiment is going on.

In a way, Edgington and Schultz are working to show Stroz that the GLM data is worthwhile. "I think the thing that surprised me the most about the National Weather Service is how much it comes down to being an art. You have all this data. But it's the person sitting in front of the computer with their experience that lets them decide whether or not to issue a warning," Edgington says. "So part of our job is to get them to try this lightning data. We try to make sure the data is good and useful so they'll want to use it and then go back to their field offices and tell everyone, 'Hey, I've got this data, it's really useful.'"

As I watch, Stroz continues to monitor the weather over Kansas. He and Schultz discuss the different data feeds—*products*, in NWS lingo—from the GLM. Stroz is accustomed to factoring lightning into his weather assessments, but until now that data has come mainly from ground-based sensors. The National Lightning Detection Network (NLDN), operated by a Finland-based company called Vaisala, provides data to not only forecasters but also insurers, utility companies, airlines, event operators, and others who need to know when lightning is coming for reasons of safety or damage assessment. Ground-based lightning sensors detect the electromagnetic energy released by lightning and use triangulation to map the location and time. Stroz has access to the NLDN and would routinely scan that along with his other displays in case of severe storms. But because the GLM uses optical sensors, it can actually "see" the discharges and provide information about the duration and brightness of lightning flashes.

As Stroz grows more comfortable with the display, he starts to see how it can aid his severe-storm forecasting. "Part of the goal here is to get me and the others out of our comfort zone," he tells me. "The goal of the experiment is to get me and the others away from using the things we're most comfortable with, our blankets, or cookies, whatever, and try to get into some of this new stuff and see if it works. We want the best stuff out in the field."

Sitting before his screens of satellite data and radar reports, Michael Stroz easily fits the image of a modern meteorologist tapping the latest in technology. But there's more to tornado safety than studying the sky. Increasingly, weather scientists have found, this work requires studying people too. A key turning point came on Sunday, May 22, 2011. That's the day the historic EF5 tornado struck Joplin, Missouri.

Joplin stands in Jasper County in southwest Missouri, just a few miles from the border of Kansas and only a few more from the northeast corner of Oklahoma. Founded after the Civil War, Joplin boomed from lead and zinc mining. These days it's known for Missouri's largest continuously flowing waterfall; the nearby George Washington Carver National Monument, which marks where the inventor was born into slavery; and the legacy of Route 66, which cut through Joplin on its way toward Tulsa.

On that sunny Sunday morning in Joplin, residents following the forecast were told to expect a high in the lower 80s and a 50 percent chance of afternoon thunderstorms. The area had been under a hazardous-weather outlook since Friday due to a forecast risk that storms could turn severe. Those who weren't headed to church might have opted to stop by a kite-flying festival getting underway that morning. For most of the people of Joplin, it was expected to be a day of celebration: The seniors of Joplin High School were getting ready to receive their diplomas. They would assemble for their graduation ceremony that afternoon at an arena on the campus of nearby Missouri Southern State University.

At 1:30 that afternoon, Joplin, along with a big chunk of southwest Missouri, was placed under a tornado watch. As activity began to build, at 3:00 p.m. the Storm Prediction Center cautioned that the "kinematic environment is very favorable for tornadoes." And by 5:00 p.m., some residents began to note rain and, in the words of one survivor, a sky that "looked very strange—like a sort of yellowish-greenish color."[30]

The tragedy began to unfold at 5:09 p.m. That's when a tornado warning was issued by the National Weather Service's Springfield, Missouri, office, which handled forecasts for Joplin. It covered western Jasper County, which included northeastern Joplin, and followed earlier tornado warnings to the west in Kansas. Two minutes later, at 5:11 p.m., local officials triggered the warning sirens for Jasper County and Joplin. The newly minted Joplin High graduates and their families were just filtering out of the event. The superintendent of schools, C. J. Huff, later told *The Joplin Globe*: "The sirens went off just as I was walking to my car." Weather Service forecasters issued another tornado warning at 5:17 p.m., this time in an area directly south of the earlier warning. It was within this box that the destruction began a few minutes later.

Investigators believe the tornado first touched the ground at 5:34 p.m., southwest of the city limits. As the twister traced an eastward path, for many it was the sound and not the sight that drew their attention. As the school band director described it to *The Joplin Globe*, it "sounded like a cross between a jet engine and a train, but the pitch didn't change." Other survivors commented on the intensity of the roaring sound as well.

Over the next roughly forty minutes, the tornado carved a trail eastward, intensifying as it moved. Officials sounded the sirens for a second time at 5:38 p.m. One survivor described the sight to researchers as a massive dark cloud; "it was an animal like coming toward me." Another, who sheltered in a small restroom in the Joplin Missouri Stake center, a local Mormon facility, recalled feeling "the roof lift off over our heads and forces lifting us skyward. Everyone's hair began standing on end [and it] seemed that a giant vacuum cleaner had its sights set on lifting us out into the stormy sky." Though the building was mostly destroyed, the occupants were lucky; they suffered only minor injuries.

Elsewhere, though, there were fatalities as the tornado flattened buildings and launched debris more than 20,000 feet into the air, according to Weather Service radar data. At some points on the tornado's twenty-two-mile journey, it reached a mile wide and its winds exceeded two hundred miles per hour. Images of the aftermath show near-total destruction: Buildings pulverized and wreckage covering the ground, interrupted only by the trunks of trees completely stripped of leaves and small branches. Joplin High School, alma mater of the new graduates, was wrecked, its walls pulled down and debris strewn everywhere.

The next day, the *Globe* reported that the official death toll late Sunday had stood at eleven but was expected to exceed one hundred as searchers recovered victims. "Teams with body bags were being dispatched to Home Depot, Wal-Mart, Academy Sports & Outdoors, Sonic and other businesses between 15th and 20th streets along Range Line Road, one of the hardest hit areas in the city," the *Globe* reported. Ultimately the number of deaths reached 161, including three indirect fatalities, with more than a thousand people injured.

Much of the story describing how this disaster unfolded is captured in the NWS's service assessment,[31] a kind of after-action report for meteorology in which experts review a major weather event, evaluate damage, and look for lessons for the future. Consistently, these reports make for compelling reading, as scientists lay out the timeline of a tragedy and dissect nature's most destructive events with a clinical eye. But even in that context, the service assessment for the Joplin tornado is a remarkable document. The scale of the storm's devastation was truly historic; no single tornado had killed more than one hundred people since 1953. The report team sought to understand why there were so many deaths and injuries in what was, in Weather Service jargon, a "warned event." In other words, the Weather Service's forecasting system had worked as designed and issued a tornado watch for southwest Missouri at 1:30 p.m., hours before the storm, and then put out tornado warnings as the actual twister spun up. Given that the tornado warning for the primary Joplin area went out at 5:17 p.m., and the twister first touched down at 5:34 p.m., the warning time amounted to seventeen minutes for those near the touchdown area and more for those farther east. "Generally speaking, advance warning of the tornado was given, information was communicated and received, and most people sought the best shelter available to them," the report stated. So how to explain the horrific outcome?

For answers, the service-assessment team looked beyond the meteorology. Team members came to Joplin and interviewed nearly a hundred people, from city officials and emergency dispatchers to actual survivors. As the report described: "The team utilized ethnographic methods or techniques commonly used by social scientists to scientifically describe cultures and the people within these cultures. In particular, the team strove to understand residents' points of views regarding the process of warning reception to warning response, and how decisions were made." At the time, this was groundbreaking.

The findings underscored the gap between meteorologists issuing warnings and the behavior of those in peril. As the report stated: "Response to severe weather warnings is a complex, non-linear process depending on perception of risk." The NWS team found that individuals digested a whole range of information when making decisions about whether, when, and where to seek shelter, from community warning sirens and TV broadcasts to social media sites and even simply eyeballing the sky. Different data points factored into the equation for different people; it depended on what they heard or saw.

Consider the story of a resident who said he was aware that storms were likely but wanted to get something to eat. While driving to a local restaurant, he heard the first warning sirens go off. When he got to the restaurant, he found that the staff had closed its doors due to the warnings and he was refused entry. Still seeking food, he drove to a second restaurant, one that was still open. He subsequently heard TV and radio reports about the tornado and was told by another customer that the tornado was in Joplin. Eventually, as the researchers put it, "management instructed protective action." The researchers concluded that this individual had been exposed to fully nine different indications of risk. One that actually made an impression was the fact that the first restaurant had stopped letting people in; he said at that point he was concerned about being out in his car. But then he received a counterbalancing signal by finding the second restaurant open. There, he was "escorted to a table and ordered a meal."

The divide is stark. At the National Weather Service, highly trained scientists used state-of-the-art technology to monitor the weather minute by minute and issue warnings. Yet one person's safety can be shaped in part by whether a restaurant manager keeps the doors open. This research serves as a reminder that humans aren't robots who automatically follow a weather warning. And it underscores the confusion possible when people attempt to process multiple forecasts and warnings along with conflicting indications, like seeing others going about their business as normal.

The sirens mentioned in the example above became a particular focus. Depending on where you live, you may not have heard the wailing tone from this type of siren. Growing up in suburban Michigan, which doesn't experience storms at the frequency of Tornado Alley but nonetheless has some every year, I knew their unmistakable sound (and that there was no need for alarm if you heard the sirens at 1:00 p.m. on the first Saturday of the month, the designated testing day). In the flat terrain of my suburb, the sound of the outdoor sirens carried everywhere. Most of these systems trace their lineage back to civil defense efforts for World War II and then the Cold War, and they do double duty, warning of severe weather as well as a possible nuclear attack. (The weather warning is a steady sound that can seem to slowly rise and fall as the siren rotates on its mount; the attack warning undulates at a faster pace.)

In tornado-prone areas, the sound of the weather siren isn't uncommon. As the NWS team found, in Joplin, it might have been too common. "The majority of Joplin residents did not immediately take protective action upon receiving a first indication of risk (usually via the local siren system), regardless of the source of the warning," the report stated. "Most first chose to further assess their risk by waiting for, actively seeking, and filtering additional information before taking protective actions." The report went on to say:

The perceived frequency of siren activation (false alarms) led a large number of participants to become desensitized or complacent to this method of warning. Many noted that they "hear sirens all the time and [sirens] go off for dark clouds," they are "bombarded with [sirens] so often that we don't pay attention," "the sirens have gone off so many times before," "sirens are sounded even for thunderstorms," and "all sirens mean is there is a little more water in the gutter."

The sirens posed an interesting problem. The meteorologists at the Weather Service aren't responsible for triggering them; they're controlled by local emergency management officials. In the case of Joplin, the report explained, the policy was to activate the sirens for a reported tornado or when storm winds were expected to exceed seventy-five miles per hour: "These triggers may or may not be associated with an NWS warning, and the Jasper County / Joplin Emergency Manager has discretion and uses professional judgment on when to activate sirens." The sirens aren't automatically activated based on the Weather Service warning but rather when the local official who is monitoring conditions—including but not limited to the NWS warnings—deems it necessary, creating the possibility of too many alerts and the related warning fatigue. The report noted this approach was "by no means unique" to Joplin. It also underscored the blunt-instrument nature of the sirens, which communicate danger but can't provide information on where people are most at risk, how severe conditions are, or what the potential impact might be. Particularly now in the smartphone era, most people can directly access more granular information. But for those who don't have access to a phone or who are working outdoors, the sirens can be a lifeline—or at least an impetus to check for more official information.

Beyond people's response to the sirens, the NWS researchers found what behavioral scientists call optimism bias. "Most individuals commented that severe weather in southwest Missouri during spring is common; however, tornadoes never affect Joplin or themselves personally," the report stated. "It was common in the interviews to hear residents refer to 'storms always blowing over and missing Joplin,' or that there seemed like there was a 'protective bubble' around Joplin." In other words, these people had come to view dangerous weather communications as the proverbial boy who cried wolf. The perception problem applied to the many other ways that residents heard about incoming weather, not necessarily official Weather Service warnings. But the findings nonetheless brought new scrutiny to the question of false alarms.

The Joplin report also found things that had gone right. There is little question that, overall, meteorologists performed well in this storm. Without the timely NWS warnings—which prompted many to take cover—the death toll would surely have been far higher. But by its nature, a service assessment focuses on opportunities for improvement. And far beyond any of the specifics, the Joplin report cemented the idea that improving tornado safety required greater recognition that better forecasts were only part of the puzzle. It marked a sharp increase in the attention paid to social science. One explicit recommendation from the report: "For future Service Assessments, NWS should plan a more structured approach to collecting information on societal aspects of warning response. This should include developing sub-teams well-versed in social science and NWS warning operations that can be quickly deployed to the field following any given severe weather disaster."

Years after the Joplin tornado, not far from the NSSL room where Michael Stroz and others at the Spring Experiment are testing out lightning data, another researcher at the National Weather Center works to unravel the societal issues. Her name is Kim Klockow McClain, and instead of studying the storms, she studies the people in jeopardy from them. Her work is crucial to doing something about the weather.

"Meteorologists get into this business because they want to save lives, they want to know that they're helping people. That's why I love being a part of this field," she tells me. "But some of the problems that are most interesting to meteorology are really about human behavior under conditions of risk and uncertainty." Klockow McClain is a social scientist who coordinates work for the national forecasting operations of the Weather Service; she previously led the NSSL's Behavioral Insights Unit. Her academic training has blended understanding of the atmosphere with analyses of human behavior: At Purdue she earned undergraduate degrees in both meteorology and economics, while her Ph.D. at OU cut across everything from atmospheric science to psychology and geography. It's her job—and that of the scientists she works with—to make sure that better weather forecasts translate to better outcomes. What makes people pay attention to a warning? What causes them to tune out? Do those at risk from specific weather dangers know the best way to protect themselves? Which subpopulations are particularly at risk and what are the most effective ways to reach them? All of these questions matter to weather forecasting even though they don't have anything to do with barometric pressure. "The atmosphere does not have free will," Klockow McClain says. "Humans have free will."

Klockow McClain is part of a new vanguard in weather forecasting: the social scientists. It's one of the most consequential developments in the battle to keep people safe from weather. Though the need to confront the human element might seem obvious, it's a relatively recent advance. Only within the past fifteen years has the role of social science gained serious traction as a vital complement to the research and technology needed to understand the atmosphere. Time and again as I learned about the impact of dangerous weather, I saw how important this work was. Factoring in human behavior is a force multiplier for advances in weather forecasting. Critically, it's also a vital tool for recognizing and addressing social inequities. Income, race, culture, language, age—all can play roles in who has a disproportionate vulnerability to different types of weather.

"For the field of meteorology, which had grown accustomed to thinking our increases in technology are making gains, 2011 was a jarring year," Klockow McClain tells me as we sit in her office at the National Weather Center. "It made everybody ask, What is it that our field actually does? Where does the rubber meet the road? And so then the social scientists are able to come in and try to suggest, here's how people actually make decisions based on what you give them."

This new appreciation informed an important tornado research project that began in 2015, VORTEX-SE. Much as the Joplin after-action report sought to understand how a well-warned event could result in so many deaths, the VORTEX-SE program focused on the question of why tornadoes in the southeastern United States seemed disproportionately fatal. Congress funded the program to be run out of NSSL, and from the start, social, behavioral, and economic perspectives were as key to the endeavor as atmospheric science and forecasting operations. The inclusion of social science was especially notable given the history of the VORTEX projects. The first VORTEX effort was launched in 1994 to study tornadogenesis (VORTEX is an acronym for "Verification of the Origins of Rotation in Tornadoes Experiment"). It featured storm-chasers in forerunners of Sean Waugh's current-day mobile-mesonet vehicle along with aircraft, all gathering data to understand the life cycle of a tornado. A successor, VORTEX2, followed in 2009. Again, the focus was on observing the storms. As NSSL described it: "VORTEX2 research teams made science history by strategically deploying all instruments on a tornadic supercell. Detailed data were collected from 20 minutes before the tornado formed until it faded away."

As the similar name might suggest, VORTEX-SE featured plenty of observations and data from mobile radars and other instruments. But it also deployed social scientists to talk to people in the affected communities. "That's one of its great strengths," Klockow McClain says. "It's really about a population. It's about the people of the southeast US. There's never been a program that's been about a particular population before." When researchers dug in, they discovered ways in which socioeconomic status played a role in tornado safety. One of these might seem obvious, but it's hugely important: the reliance of lower-income individuals and families in the region on mobile homes, which are typically less able to stand up to high winds than permanent-structure housing. (Modern mobile home–style units are often referred to as "manufactured housing," which is something of an upscale rebranding. Think of a "used car" versus a "certified pre-owned automobile.") As a VORTEX-SE overview report explained: "Mobile and manufactured homes have the highest fatality rate of any housing structure, with estimates of fatality rates of 15 to 30 times that of houses with roofs and walls tied to the foundations."[32]

The surveys also found something less obvious: Mobile-home residents, like most people, were aware of the general guidance for sheltering in a tornado: Get to an interior room on the lowest floor, ideally the basement. But that guidance doesn't apply to mobile homes. The safety recommendation for those living in mobile homes is to get to a designated shelter (some mobile-home communities have these) or a permanent structure. "They've gotten our big safety message," Klockow McClain says. "But if you're in a mobile home, it just doesn't work. You have to get out of there." The problem is more complicated than informing people of the appropriate way to shelter. For someone in a traditional-construction home or an indoor workplace, moving to a safe shelter requires only a few minutes. But for mobile-home residents, the nine or so minutes after a tornado warning may not be enough for them to get to a safe site.

The VORTEX-SE project turned up other issues too, from the difficulties faced by those with disabilities who might not be able to hear or see warning broadcasts to the challenge of tornadoes that form during the night. As the overview report summarized: "Results reveal that 84% of participants who were asked about a daytime tornado warning self-reported that there was a high or very high chance they would find out about it, whereas only 48% of participants indicated as such when asked about a nighttime warning." And though the integration of social research was a pioneering aspect of VORTEX-SE, there were also important meteorological findings. The typical tornado forms in a supercell, but others are spawned from those squall-line storms known to meteorologists as quasi-linear convective systems. Unlike the individual supercell, a QLCS is a family of storms stretched out over a line that can extend hundreds of miles. Supercells are more strongly associated with tornadoes, particularly big ones; QLCSs are more known for strong but non-tornadic winds and heavy rain. They can produce tornadoes, although the twisters are usually less intense than those from supercells. But the southeastern United States experiences a disproportionate number of QLCS tornadoes, and the VORTEX-SE work pointed to a need for better forecasting capabilities with these storms. Each single finding is, of course, important on its own. But the VORTEX-SE approach fit all these pieces together for a broader understanding of tornado risks in the region. It wasn't long before the people of the Southeast suffered a painful example of the factors at issue in the VORTEX-SE work.

On Sunday, March 3, 2019—well before the height of tornado season—dark clouds gathered over the Southeast and resulted in an outbreak of tornadoes. NOAA later assessed a total of seventy confirmed tornadoes that day spanning Alabama, Georgia, South Carolina, and the Florida

Panhandle. Most did relatively minor damage. One did not. Around two o'clock in the afternoon local time, a tornado touched down in southeast Alabama. Beginning in the northeast part of Macon County, the tornado began to strengthen and moved on a track continuing toward the northeast. Over the next twenty-nine minutes, it ripped across that portion of Alabama, crossing into Lee County. It grew into an EF4 tornado with winds of 170 miles per hour; it tracked across Lee County, through the town of Beauregard, and eventually into nearby Smiths Station before dissipating.

That single tornado killed twenty-three people. Homes were destroyed, automobiles flung like toys. Near the intersection of Lee County Roads 36 and 39, the Weather Service found, "The tornado bent the frame of a car around the remnant of a large tree whose upper portion had broken off and totaled three vehicles by severe impacts into the bases of two remaining tree stubs."[33] David McBride, the owner of the Buck Wild Saloon near Smiths Station, told the *Opelika-Auburn News* about his narrow escape. Noticing trash swirling in the air, he drove toward a nearby gas station with the intention of sheltering there. "I'm a really, really lucky man to be here because had I sat there for another 10–15 seconds, it would have thrown my car across the highway," McBride told the newspaper. The twister ripped the roof off the Buck Wild Saloon and scattered chairs from the bar along the highway, the *Opelika-Auburn News* reported.

Warning time, unsurprisingly, was a focus in the aftermath. The tornado warning for southern Lee County, including Beauregard and Smiths Station, was issued by the National Weather Service office in Birmingham, Alabama, at 1:58 p.m. local time, a few minutes before the twister's initial touchdown. Later estimates put the tornado's arrival in Beauregard around 2:07 p.m., meaning people there had at most nine minutes of warning. (Since many people receive warnings funneled through local TV or radio rather than directly from the Weather Service, the time window might have been even slimmer.) But the official tornado warning is only the last line of defense, and in the hours and even days leading up to the tragedy, forecasters were pointing to the potential for highly dangerous weather in the region for that Sunday.

Back in Norman, Oklahoma, at the National Weather Center building, meteorologists at the Storm Prediction Center pore over computer models to identify severe thunderstorm and tornado risks. These SPC forecasts represent a critical heads-up. On Thursday, February 28, 2019, the SPC's severe-weather outlook drew a yellow blob around much of Mississippi and about two-thirds of Alabama and labeled the risk *15 percent*. Though that might sound low, if you spend much time with SPC outlooks, you learn to pay attention to any projection of severe weather.[34] By the next day, Friday, March 1, the outlook for March 3 focused the area of highest risk more squarely on the southern two-thirds of Mississippi and Alabama, including Lee County. The risk for severe thunderstorms in that area was labeled *Slight*, with a larger circle around the area designating the risk as *Marginal*. Again, in the context of SPC outlooks, that's a risk that shouldn't be taken lightly. On Sunday, March 3, the same-day convective outlook placed a smaller orange oval around an area encompassing Lee County and on into Georgia, with the risk noted as a more alarming *Enhanced*. The SPC summary stated: "Severe thunderstorms with damaging gusts and tornadoes are expected over parts of the Southeast today."

By late Sunday morning, the SPC felt conditions warranted a tornado watch. (Under the NWS system, watches are issued by the SPC, while actual warnings are handled by the local weather forecast office.) A watch bulletin issued at 11:40 a.m. local time, in effect until 6:00 p.m., stated: "Thunderstorms will continue to intensify this afternoon ahead of a strong cold front. Supercells will be possible along and ahead of the front, capable of damaging wind gusts and a few tornadoes. Strong tornadoes are possible." The watch included this boilerplate language: "REMEMBER . . . A Tornado Watch means conditions are favorable for tornadoes and severe thunderstorms in and close to the watch area. Persons in these areas should be on the lookout for threatening weather conditions and listen for later statements and possible warnings." So even though the tornado warning gave residents only a scant nine minutes to take shelter, the danger signs had ramped up steadily beginning a few days earlier, first with the SPC outlooks, then the tornado watch. But many residents might not have taken those early cautions seriously, assuming they even followed weather forecasts closely enough to know about them.

Lee County is not a wealthy place; the median household income is just under $60,000, around three-quarters of the US median. About ten minutes south of Opelika, the county seat, stands Beauregard, named for a prominent Confederate army general. Given the economics of the area, it is perhaps not surprising that mobile homes—one of the factors scrutinized by the VORTEX-SE work—were common. The consequences were deadly. Of the twenty-three fatalities from the Lee County EF4 tornado, nineteen occurred in manufactured homes, according to a team from StEER, the Structural Extreme Events Reconnaissance Network.[35] Of the remaining four deaths, three occurred in a modular home (like mobile or manufactured homes, modular homes are prefabricated, but they're typically placed on a traditional foundation) and one in a standard-construction, or site-built, home. The StEER report found: "Lack of proper anchorage in both older site-built homes, and manufactured homes of all ages, appears to be a key contributor to the enhanced fatality rates" that were observed in the EF4 Beauregard / Smiths Station tornado.

Viewed through the lens of the VORTEX-SE work, the contours of the Lee County disaster come into better focus. Extra minutes of lead time for the actual tornado warning have obvious value. But the particulars of the community mattered. The vulnerability of people living in lower-cost mobile and manufactured homes contributed to the deadliness of the event. As the StEER analysis noted, "While sheltering in a manufactured home is not recommended, many residents are limited in their sheltering options, particularly in rural settings when the distance to the nearest community shelter is prohibitive."

Two days after the EF4 in Alabama, I spoke with Erik Rasmussen at NSSL. Rasmussen is a veteran storm-chaser and one of the most experienced tornado researchers around, having worked on the original VORTEX project and more; for VORTEX-SE, he served as coordinating scientist. I had been watching the Storm Prediction Center outlooks with interest in the days leading up to the March 3 tornadoes and noticed not only the timely issuance of the tornado watch that included Lee County but also the strident tone of guidance. Shortly after the tornado warning, the Birmingham WFO stated on Twitter: "TORNADO EMERGENCY for southern Lee & northern Russell Counties! Large & EXTREMELY DANGEROUS TORNADO ON THE GROUND near Dupree, moving East. PLEASE TAKE SHELTER NOW if you live between Dupree & Smiths Station!!" Rasmussen and his VORTEX-SE colleagues had been monitoring the developing weather closely, watching computer models and, as the storms began, radar and data from sounding balloons. I put it to him directly: How can we be so good at predicting dangerous weather and still have so many people die? For Rasmussen, that dichotomy demands progress on all fronts—forecasting, warnings, and understanding human behavior and community vulnerabilities.

"I think the outlooks and warnings were very good," Rasmussen told me. "I'm beginning to sense there's a small gap in time, though. The warnings are great, ten to twenty minutes. The outlooks are really good at a few hours. But I'm beginning to wonder if maybe we can do better in the gap between those two." He explained that as evidence of the threat increases in that middle time period, that's an opportunity for people to take lifesaving action. If a mobile-home resident in a rural area doesn't have a safe shelter in the immediate vicinity, ten minutes of warning may not offer any real choices. But prior to the warning—especially with a plan arranged in advance—there would be time to reach or at least move closer to a safer place. "We have to help people to identify survivable structures that they can get to in a period of time that's appropriate," Rasmussen said. "You can't run to a nearby town or a nearby survivable structure with a minute or two notice. Your options become that you basically don't have options. Running to the interior room in the lowest level doesn't apply to a structure that is going to be destroyed."

Saving more lives from tornadoes represents one of the weather world's most demanding problems because it involves so many factors, from the science and technology of observing the atmosphere to the societal challenges of making sure that people are prepared and ready to act in time.

One promising approach takes advantage of new capabilities of computer weather models but also requires disrupting decades of communication about tornadoes. It's called the Warn-on-Forecast System (WoFS). To understand how big a shift is involved, remember that tornado warnings are issued based on observing evidence of a twister actually forming, what's known as a warn-on-detection approach. The goal of WoFS (pronounced "*wah*-fs") is to introduce better information between a tornado watch and a tornado warning, much as Erik Rasmussen described after the Lee County tornado. Patrick Burke, the program lead for WoFS at NSSL, walks me through the advancements that have helped drive its development. One key to WoFS is a category of computer weather model called a convection-allowing model, or CAM. Usually when the public hears about numerical weather-prediction models, the discussion centers on hurricane forecasting. Those models—which include the massive Global Forecast System run by NOAA and the Euro model from the European Centre for Medium-Range Weather Forecasts—simulate the entire Earth atmosphere days in advance. CAMs operate on a much more granular scale in both time and geography, essentially holding up a magnifying glass to areas where convective storms will form and updating the predictions frequently. "The science got better and better at modeling these individual thunderstorms that are only a few miles across," Burke explains. "And we began to see an opportunity there."

That opportunity: Issue alerts that have greater certainty than the forecasts driving tornado watches but still in advance of a tornado warning based on detection. "On average, we're giving about thirteen minutes for people to receive a warning, understand it, react to it, and go to their safe spot," Burke says. "That might work for a family who's sitting at home and knows they have a basement to go to. But there are so many vulnerable people out there who need more time, whether it's a trailer park, or hospitals who might need to move a lot of people who might not be mobile, or airports, or sporting venues."

A case study from May 16, 2017, illustrates the potential. With tornadoes likely that day, meteorologists in the Norman WFO—who work side by side with the SPC and NSSL in the National Weather Center—collaborated with WoFS researchers. A tornado watch was issued for the Texas Panhandle and western Oklahoma at 1:50 p.m. local time. Then, with WoFS guidance showing tornadoes likely, forecasters issued an unusual significant weather advisory at 5:45 p.m. stating: "Severe weather is likely with these storms as they move into Oklahoma and there is a high probability that tornado warnings will be issued."[36] Twenty-eight minutes later, at 6:13 p.m., the Norman WFO forecasters issued the first tornado warning for a storm cell headed to Elk City, Oklahoma. Subsequent tornado warnings (at 6:45 p.m. and 7:08 p.m.) included Elk City itself.

The WoFS predictions were correct. Elk City was struck by a tornado later evaluated as an EF2. A subsequent NWS survey found that the tornado killed one person in a tragedy that seemed to encapsulate the challenges of tornado safety: "According to law enforcement, the victim had been in a mobile home, then left in his vehicle to try and find more substantial shelter. He had returned to his mobile home and it's believed he was still in the vehicle when the tornado hit." Local news on KWTV, also citing law enforcement, reported that the tornado lifted the victim's vehicle and threw it several hundred feet. Damage assessments concluded that sixty-two homes in Elk City had been destroyed.

Reviewing the timeline shows that the "high probability that tornado warnings will be issued" notice provided guidance with a lead time anywhere from twenty-eight minutes for those in the first warning area, southwest of Elk City, to almost an hour for those in the more densely populated Elk City area. Afterward, an Elk City official said, "Based on the information from the NWS we were able to activate outdoor warning sirens about 30 minutes ahead of the tornado," according to an NSSL account. The Weather Service continues to develop WoFS to ready it for routine operational use. That will involve not only improving the model guidance but getting that guidance to a point where meteorologists can feel consistently confident in making a pre–tornado warning announcement.

It also means making sure that key consumers of tornado information are ready to receive this kind of bulletin. The traditional tornado warning is what meteorologists refer to as a deterministic forecast. It's binary: Yes or no; the tornado is coming or it's not. WoFS works instead with a probabilistic forecast. That special advisory for May 16 didn't *guarantee* a tornado. It told of a "high probability." Will people, even experienced officials like emergency managers, know what to do with that? "We're trying, as a community, to get better about communicating what we do know and what we don't know," Burke says. "A lot of times we frame that as quantifying uncertainty. But in a warn-on-detection paradigm, like we've been operating in for decades, that expression of uncertainty isn't there at all, right? So it's going to take a lot of learning, and it's not only a cultural shift on the part of the forecasters but also work with the end users." Thanks to social media, the Weather Service can more easily communicate information directly to the public. Realistically, though, this transformation will depend on educating actors throughout the weather world, from emergency managers and local officials to the broadcast meteorologists who give local forecasts. NOAA and the NWS move carefully on any changes to warning communications, studying them for unexpected consequences and proceeding slowly to avoid confusion that could prompt people to tune out of the forecasts.

Humans, unfortunately, are notoriously bad at processing probabilistic information. The brain seems to crave certainty. Tell someone that there's a 65 percent chance of something—whether it's rain tomorrow or the probability of a candidate winning an election—and there's a tendency to believe that's what will happen, even though there's a one-in-three chance it won't. Before we ask people to make high-stakes decisions about weather forecasts such as evacuating their mobile home or clearing an outdoor stadium, we need to make sure we've effectively communicated the uncertainty that Burke describes along with the value of getting prepared for an event even though it might not come.

Research has found some factors that can encourage attention to warnings. Clear information about affected locations helps people determine whether they're in the danger zone. Impact-based descriptions—that is, warnings that describe not just the weather threat but also what could happen—also help. For instance, a flood warning for an urban area might communicate that staying in a basement apartment risks death by drowning.

But because people receive and process information in so many ways and from so many sources, the Weather Service can't devise a magic-bullet solution. Indeed, having a message reinforced by multiple sources, especially trusted ones, helps drive the peril home. Before the digital age, those trusted communicators were primarily television and radio broadcasters. Now many people, particularly younger individuals, get information from social media. That presents risks and opportunities. The risk is the potential for low-quality information or downright misinformation. The opportunity is directing people to community leaders, influencers, and friends who can reinforce the need to take action.

What else can be done for specific vulnerable groups? Mobile-home residents need suitable shelters, even if they must be built. Organizations such as hospitals and schools typically have response plans for tornado warnings, but safety officials can help those run more smoothly by paying attention to longer-range outlooks and ramping up readiness before an actual warning. To reach non-English-speaking communities, the National Weather Service can expand its communications beyond English and Spanish—though local news sources and even neighborhood leaders can help by highlighting severe-weather forecasts and providing information about how to respond in an actual warning. Such education efforts can be especially helpful for threats outside of Tornado Alley. Individuals can prepare by paying attention to watches and—for the danger of nighttime storms—by purchasing inexpensive weather radios that will click on when warnings are issued. They can also improve their weather literacy by making sure they understand the meaning and implications of terms like *watches* and *warnings*.

As Sean Waugh navigates the Probe 2 truck on the rural roads of western Oklahoma, I get to watch just how quickly the sky can go from mostly clear to apocalyptically dark. Waugh and Kiel Ortega monitor a number of data feeds besides the display from Probe 2's own instruments. An important one is the radar image of the area from the NEXRAD network. They walk me through how to view this data myself by downloading the RadarScope app to my iPhone. (There are a variety of apps that take the output from the government-run NEXRAD network and display it, but RadarScope, I learned over time, is the go-to for many pros, and once I began to understand what I was seeing, it became addictive.) They also check the forecasts being issued by the SPC back in Norman, including a mesoscale discussion, an inside-baseball description of conditions in areas likely to experience severe weather. Waugh and Ortega weigh all of this along with their own experience to decide which direction to head to find a storm getting underway. As we steer to the next location of interest, they talk about why they're so committed to storm-chasing.

"For me, every time we go out, you see something you've never seen before," Waugh says. "It's a constantly changing environment, and we don't fully understand it. You can read a textbook, you can read papers, you can talk to people about how the atmosphere works. But when you actually go out and watch it, you see it unfold right in front of you. When you see all those pieces come together, it kind of makes everything click. It's a challenge, because you're literally trying to predict the future." Ortega adds: "It's a chess game." The storm-chaser, he says, must think several moves ahead, hoping to get the truck to the right spot.

We stop several times to launch weather balloons. The process can be almost comical, sometimes requiring the researchers to wrangle the balloon against whipping winds while it finishes inflating. Once aloft, the instrument package transmits data on temperature, pressure, and humidity. The readout in Probe 2 shows this on a graph known as a skew-T chart, which plots the temperature and dew point. These observations are called a sounding of the atmosphere. Across the United States, local Weather Service offices send balloons up twice a day. The soundings from Probe 2 supplement those. "We want to actually get out into that region where the Weather Service doesn't have any access and where the storms are going to be," Waugh says. The data gets sent back to Norman and the Storm Prediction Center. "It gives them access to information they don't normally have, which on days like today is really critical."

This day's chase began around 9:30. As Waugh explains, starting early gives a chance to get out and take measurements as conditions evolve toward severe storms. But starting later means the likely locations for tornado-producing storms are more clear, so the chase can deploy to an area with higher confidence. As we hit the afternoon hours, the storms begin to heat up. Several times we find ourselves pummeled by hail or confronted by the eerie spectacle of a wall cloud dipping down over farmland dotted by wind turbines. In one, I see a funnel shape forming and the rotating tube begins to tilt toward the ground before it peters out.

Despite the intense weather, I never feel a sense of danger. Waugh and Ortega are scrupulously careful about safety. At one point, after running into a storm-chaser traffic jam, they decide to reverse direction. At this time of year on the Great Plains, the roads can become overrun with professional and amateur chasers, some doing research, some looking to capture photos and videos they can sell, some just hoping to see a tornado in real life. We encounter an armor-plated chaser vehicle that looks like someone combined a stealth fighter jet with a van—all flat black surfaces and angles, including a skirt that extends nearly all the way to the road surface. Tornado season has always attracted chasers, but after the popularity of *Twister*, you can now find Tornado Alley tours advertised on the internet. Waugh and Ortega prefer to avoid the crowds; if a dangerous storm shifts, they want an easy and clear path out of the area.

In late afternoon, we start making our way back east to Norman. It's been a long day. We've been on the road for over ten hours. As we pull into the garage of the National Weather Center, we hear a wailing sound. The local emergency sirens are going off; a tornado warning has been issued for Norman itself. It's odd to experience a warning while inside the very place from which forecasters issued it. When we extract ourselves from Probe 2, we learn the building is on lockdown due to the tornado warning. I'm directed to the National Weather Center auditorium, which was built belowground to serve as a shelter. I'm grateful because, as a visitor to the area, I would have had no idea where to go for shelter—a problem for anyone away from home. After about forty minutes, the all-clear sounds and everyone files out of the auditorium. A damage survey later confirms a tornado passed through the area, at one point reaching EF1 winds of sixty to seventy miles per hour. One home was damaged, as were a bunch of trees. No one was injured.

2 Fire

WATCHING THE WIND, STOPPING THE SPARKS

The appeal of San Diego isn't hard to figure out. The sun shines, the Pacific Ocean gleams, and the vibes are laid-back, even by Southern California's standards. Looking out over the Marina District, I see the masts of sailboats on the bay juxtaposed with the skyline's high-rise apartments and condos. Scenic beaches dot the long, gorgeous coastline, and multimillion-dollar homes gift their owners with tranquil ocean views. Apart from a few wetter months in the winter, the weather is mild and sunny and rarely disappoints.

If you head inland, the landscape changes with remarkable speed. Driving east from downtown, I watch the waterfront and city streets drop away. I pass through suburban communities and soon reach the foothills, with the Cuyamaca Mountains rising up, and the Laguna Mountains beyond them. The hills wear a mottled coat of low, shrubby trees and plants. You don't need to be a botanist to recognize that this land is more arid than lush; dry-looking sandy patches of dirt fill the spaces between the sprawling brush.

This is wildfire country, and the weather here matters a lot. The amount of moisture from rain, the strength and direction of the wind, the possibility of a spark from a storm's lightning—all are variables that can mean the difference between a normal sunny Southern California day and a raging blaze that puts lives and homes at risk. In some locations, the ecosystem might be considered chaparral, a plant community dominated by undergrowth such as chamise, a flowering evergreen shrub, according to the California Native Plant Society, a nonprofit conservation group.[37] Other areas might be deemed coastal sage scrub ecosystems, with lots of California sagebrush, a shrubby deciduous plant. This vegetation is quite flammable when dry. In fact, as with so many forest and plant ecosystems, nature has woven fire into the region's life cycle. Dormant seeds from chamise shrubs, for instance, germinate after a fire cracks open their outer layer.[38] Fire is a component of the plant's survival. The problem comes when these plant habitats butt up against the places where humans want to build, live, and work. That's what the fire people refer to as a wildland-urban interface, or WUI.

San Diego County, like so much of California, has lots of WUIs. Suburbs sprawl; people want backyards and scenic views. Part of the price they pay for that is living with the risk of wildfires. The worst to strike the area happened in 2003. It was called the Cedar Fire.[39] The cause, according to officials, was a signal fire lit by a hunter lost in the forest.[40] From that humble source, the Cedar Fire spread and raged for weeks, ultimately destroying over 2,800 homes and other structures. It took the lives of fifteen people, including a firefighter. Ultimately the blaze burned over 273,000 acres. For perspective, that's an area nearly five times the size of the city of Boston. Though the Cedar Fire was San Diego County's worst fire, it was far from the only one. There have been several dozen since the start of the twenty-first century. In 2020, there was the Valley Fire, which burned 16,390 acres and wiped out twenty-four homes but fortunately killed no one.[41] Like so many regions in the western United States, people here live with the possibility of wildfire in a way that's difficult for those elsewhere to understand.

I have come to San Diego to observe a state-of-the-art weather-forecasting operation crucial to the efforts to prevent wildfires. It's led by a veteran meteorologist who has built a program that is the envy of many weather experts. In its high-tech operations center, which evokes NASA's Mission Control, there are giant screens at the front of the room and more than two dozen workstations. The workstations are arranged in three rising tiers of stadium-style seats to ensure that those in the back can clearly see what's going on up front. A sign hangs from the ceiling above every workstation identifying its function. Here, information pours in from weather sensors throughout the region, from the border with Mexico all the way up to Orange County. A forecaster can monitor everything from basic temperature and humidity to the legendary Santa Ana winds that can turn a tiny fire into a rampaging blaze. Data feeds into a customized computer-forecast model, and artificial intelligence helps sharpen the model's skill.

This isn't a National Weather Service office or some other NOAA facility or any of the federal, state, or local government agencies tasked with wildfire response. Instead, I'm at the electric company. Specifically, I'm at the emergency operations center of San Diego Gas & Electric (SDG&E). This utility provides electricity and gas to some 3.7 million people in San Diego County and the southern part of the abutting Orange County. The wires that carry SDG&E's electricity crisscross that region from the city of San Diego to those shrub-covered hills inland. And because a spark from those lines can trigger a wildfire disaster on a dry and windy day, SDG&E cares very much about the weather. With climate change contributing to hot weather that sucks the moisture from vegetation that's already dry for want of rain, ramping up the fire risk, the people at SDG&E care more all the time. "We didn't realize it ten years ago. We thought we were just dealing with mitigating wildfires," says Brian D'Agostino, SDG&E's vice president of Wildfire and Climate Science and the meteorologist who presides over this ambitious forecasting center. "But now, after a decade, you realize that we're writing a blueprint for climate adaptation and how you start dealing with this increased amount of extreme weather. Our company has been around for a hundred and thirty years, but in the last ten years things are changing drastically. So we're adapting quickly."

The in-house meteorology operation at SDG&E is innovative enough that it merits attention on its own. I've been fascinated by this kind of private forecasting because it receives so little notice; most people aren't aware of it. But I've also come to San Diego to learn how SDG&E's work fits into a larger ecosystem of weather forecasting and fire response for the area. That ecosystem spans companies, local, state, and federal agencies, and even nonprofits. It contends with factors that can be controlled, such as human-caused ignition sources, and ones that can't, such as dangerous winds. Members of this ecosystem work to prevent fires, predict their behavior when they do occur, and warn and support people who will be in harm's way. It offers a powerful model of the innovation and cooperation needed to protect against the threats from extreme weather.

Imagine facing an incandescent wall of flames three stories tall or watching a long line of fire creeping across the countryside toward you, your home, and your family. A wildfire represents a uniquely terrifying threat among nature's many environmental dangers. As destructive as they may be, tornadoes quickly dissipate, floodwaters subside, and hurricanes move along on their track and eventually peter out. But under the right conditions, a wildfire multiplies, infecting broad expanses of land like a virus spreading through a population. Wildfires can burn at 2,000 degrees Fahrenheit, easily hot enough to melt aluminum. Some forest fires move as fast as six miles per hour or more, meaning you would need to keep up a healthy jogging speed, potentially over difficult terrain, to outrun them. Their plumes of smoke, which can sometimes be seen from orbit, travel thousands of miles and alter the colors of the sunset on the other side of a continent. Wildfires destroy homes and businesses, force evacuations, trap people in dangerous circumstances, and put firefighters at risk. Once they've started, they are difficult and expensive to stop.

Strictly speaking, wildfires may not be a form of weather. But weather and fires are deeply connected. Dry conditions make plants and trees more likely to burn. Lightning can spark a blaze. Winds can feed and spread flames, turning a small fire into a dangerous inferno. Fire-weather forecasts are critical for safety. Just as people living in the Great Plains are accustomed to hearing storm outlooks and tornado warnings in their forecasts, residents of California and other western states expect their weather reports to include a red-flag warning when conditions raise the risk of wildfires. For the public, fire-weather forecasts and red-flag warnings advise caution around open flames and other potential ignition sources. Beyond staying vigilant and getting prepared to evacuate in case of an actual fire, individuals can do little else.

For an electric utility, with those power lines stretched far and wide near combustible shrubs and trees, it's a different story. The calculus of operating during fire weather is far more complicated, and the need for accurate forecasts is critical. From airlines planning their operations to commodities traders placing bets on crop yields, many industries have their own specialized needs for meteorology, and weather intelligence can be enormously valuable. Much of this forecasting work takes place out of the public eye. In the case of fire weather, NOAA and the National Weather Service play an important role, but companies like SDG&E want even more information in the form of customized and hyperlocal forecasts that can guide their decisions. Fire danger now represents one of the most significant areas for cutting-edge private forecasting, both for preventing conflagrations and for fighting them effectively.

The stakes keep growing. Aggravated by a combination of seemingly inexorable construction of housing near fire-prone areas and the effects of climate change, the dangers that D'Agostino and his colleagues worry about worsen each year. First, consider the vast reach of housing developments. Volker Radeloff, professor of forest and wildlife ecology at the University of Wisconsin–Madison, and other colleagues at the university's SILVIS Lab, which is focused on conservation and sustainability, warned about the rapid expansion of homes in a 2018 paper.[42] They found that the WUI—that problematic wildland-urban interface, regions where housing and other construction borders undeveloped land—grew at an alarming clip in the United States from 1990 to 2010. The number of new houses in US WUI areas jumped from 30.8 million to 43.4 million. WUI land increased 33 percent to 770,000 square kilometers, an expansion of 189,000 square kilometers or about 73,000 square miles, an area bigger than North Dakota.

Next, add in the changing climate. Scientists studying the link between climate and fire say that as the air in wildfire-prone regions becomes warmer and drier, it pulls moisture out of vegetation. Those desiccated plants then ignite and burn more readily.[43]

With climate helping to create fire-friendly conditions and development putting more homes in harm's way, the twenty-first century is already racking up troubling evidence of the growing threat. The California Department of Forestry and Fire Protection, known more colloquially as Cal Fire, keeps a running list of major blazes. Of the twenty largest wildfires on that list as of May 2024, fourteen have occurred since 2010.[44] All but two have happened since 2000. The deadliest fire in California history, the 2018 Camp Fire that raged north of Sacramento, claimed eighty-five lives. Of course, the threat isn't confined to the Golden State. In 2020, the Cameron Peak Fire became Colorado's largest ever, with more than two hundred thousand acres burned but fortunately no fatalities. Throughout the western United States, communities must gird themselves each year for the potential of deadly infernos nearby. The economic impact can be enormous. NOAA estimates the 2018 wildfire season in the western United States cost $30 billion; for the year prior, that number was $23 billion.[45]

Better weather forecasting can help. From week-ahead predictions for dry and windy conditions that can guide community preparations to short-term, high-resolution models that pinpoint areas of concern, knowing more about the weather on the way creates opportunities to prevent fires. I spoke with experts at the National Weather Service who track conditions in fire-prone areas and researchers who are bringing new high-powered computing techniques and artificial intelligence to these forecasts. But to really understand the unique challenges of fire weather, I talked with the meteorologists behind the scenes at companies developing their own tailored solutions.

Electric utilities are uniquely positioned to put these forecasts to use, pulling out all the stops to avoid a spark that could become the next statistic in Cal Fire's lists of destructive fires. While state and local agencies can urge the public to be careful in their activities and behavior, utility companies can inspect equipment, cut down tree branches threatening lines, and, most drastic, shut off power in specific areas when the danger level spikes. But particularly in the case of power shutoffs, the trade-offs can be vexing. Turning off the juice eliminates a potential ignition source but also creates, at the very least, inconvenience for residents and businesses in the affected areas and, at the worst, dangers for vulnerable individuals who rely on power for medical devices, such as home dialysis machines. As meteorologists get better at predicting conditions, utilities hope to restrict such extreme measures to the smallest footprint possible and the least time necessary.

Fire helped make our civilization possible. Our ancestors relied on it as a source of heat, a way to repel predators, and a means of cooking food. But fire is, as the saying goes, a good servant but a bad master. Modern civilization, through climate change and development in fire-prone areas, keeps making fire a greater threat. To confront it, we need to see the wind coming.

When residents of the town of Paradise, California, woke up on Thursday, November 8, 2018, the morning seemed ordinary. The *Enterprise-Record*, the daily newspaper in nearby Chico, told of lingering issues from Tuesday's Election Day, with thousands of provisional and mail-in ballots in Butte County remaining to be counted. The paper, known locally as the *E-R*, also updated readers on the two-week-long Farm City Celebration bus tour promoting local agriculture. Paradise is nestled in the foothills of the western edge of the Sierra Nevada Mountains, about ninety miles north of Sacramento and a hundred and forty miles northeast of San Francisco. Each year in late April, the town's Gold Nugget Museum hosted an event celebrating the 1859 discovery of a chunk of gold weighing over fifty pounds, complete with the annual naming of a Miss Gold Nugget. Based on the 2010 census, more than twenty-six thousand people called Paradise home.

The potential for an extraordinary fire event had been building for a few days. Along with a wide portion of the region, Paradise had been experiencing warm and dry weather, with highs in the high 60s to low 70s. On Monday, the Sacramento weather forecast office of the NWS had declared a fire-weather watch for Wednesday through Friday, with gusting winds and humidity during the day dropping to a desert-worthy 10 to 15 percent. One day later, the Sacramento WFO upgraded its guidance to a red-flag warning. "Please practice fire safety!" the forecasters urged on social media. By Wednesday, with winds in the foothills predicted to pick up and gusts from forty to forty-five miles per hour expected, the national forecast from the Weather Service's Storm Prediction Center declared the area's fire-weather outlook for Thursday was "critical."

The trigger for disaster struck before dawn on Thursday. At 6:15, the grid control center for the Pacific Gas & Electric Company (PG&E) noted an "interruption" on a 115,000-volt electric transmission line in the Feather River Canyon east of Paradise.[46] A component called a C hook had failed, according to a subsequent investigation. The C hook helped separate a high-voltage line and its insulator from the steel tower suspending the power lines high above the ground. When it gave way, the hot line touched the tower, resulting in an arc of electricity so sizzling that it melted aluminum in the power line and steel in the tower. Aluminum melts at 1,221 degrees Fahrenheit and steel at 2,600 to 2,800 degrees, depending on its composition. When the molten metals dripped onto dry brush below, their heat was more than enough to ignite a fire. A few minutes later, a PG&E employee driving nearby spotted what looked like a fire at the base of a tower, according to the investigation later released by Butte County officials.

Earlier that morning, fire captain Matt McKenzie had awakened in Cal Fire's Station 36 to the sound of what's known as the Jarbo winds, named for nearby Jarbo Gap. According to the investigation, McKenzie heard a rain-like sound on the station's roof while preparing breakfast for the firefighters. The sound turned out to be pine needles blown off trees and striking the building hard—an unsettling indication of how much the winds were picking up. Station 36 was located east of Paradise and about seven miles southwest of Pulga.

The firefighters' breakfast was interrupted by the report of a possible fire. The PG&E employee's sighting had been quickly relayed to emergency operators, and at 6:35 a.m., two fire engines were rolling from Station 36. At this point, events were already unfolding at a terrifying pace. At 6:44 a.m., once McKenzie got eyes on the blaze, evaluated the terrain and wind, and assessed the options, he "concluded there was no available route to attack the fire," according to the Butte County investigative summary. McKenzie decided that residents of Pulga, a town between the fire and Paradise, should be evacuated. The radio crackled with McKenzie's sobering report: "This has got potential for a major incident."

The fire was designated as the Camp Fire, taking its name from Camp Creek Road near the outbreak site. By 7:44 a.m., a little more than an hour after the initial spark, the Camp Fire had raced through the towns of Pulga and Concow. The flames were now at the edge of Paradise. The wildfire had traveled some seven miles in an hour. Residents scrambled to flee, resulting in gridlock on the roads as the danger bore down on the town. Staff at the Feather River Hospital in Paradise frantically loaded patients into their own cars and trucks to evacuate. A nurse at the hospital described the scene to the *Chico E-R*. First, they got word that a fire was seven miles away. "Fifteen minutes later, it was on the grounds," he said.

It took eighteen days, until Sunday, November 25, for the Camp Fire to be "fully contained," in the parlance of firefighting—a phrase that doesn't indicate that nothing is burning; it means that firefighters have encircled the blaze with a perimeter, preventing further spread. But even after one day, the scale of the catastrophe had begun to reveal itself. In Paradise, homes had been eradicated. The Fosters Freeze burgers and ice cream shop was gone. So, the *E-R* reported, was the Paradise Elementary School. The Gold Nugget Museum? Destroyed. Burned-out cars lined the roads. Beneath some of the charred automobiles was a bizarre sight: bright silver lines tracing irregular paths on the ground like rivers inked on a map. The heat of the fire had melted the aluminum in the cars' wheels.

Of course, the ultimate tragedies were the lives lost. In the days after the fire struck Paradise, the death toll kept rising. In total, the Camp Fire took eighty-five lives, according to a tally by authorities in 2020. The Butte County district attorney's investigative summary contains a heartbreaking list of individual victims. Some, like Cheryl Brown, seventy-five, and her husband, Larry Brown, seventy-two, of Paradise, died in their homes. Ms. Brown was found seated in a recliner next to her husband, the report stated. Others, like thirty-six-year-old Andrew Burt, died in their vehicles. "Based upon the evidence, Mr. Burt had been in the minivan attempting to escape the fire when the minivan was overcome by the fire," the report said. "There were three other vehicles containing the remains of four other victims near the minivan." One man was found outside his home, ten feet from the wheelchair he relied on. Investigators concluded he "tried to escape the flames by dragging himself along the ground."

The Camp Fire would prove to be the deadliest wildfire in California history. Many lives were lost, and many others were disrupted; according to the investigation, Cal Fire concluded that the Camp Fire destroyed "13,696 single-family residences, 276 multi-family residences, 528 commercial residences, and 4,293 other structures." It spurred an extensive investigation that involved Cal Fire arson experts, evidence specialists from the Federal Bureau of Investigation, prosecutors from the California attorney general's office, and the Butte County district attorney's office. Four months after the fire, a special grand jury began reviewing evidence, focusing in part on how the fire was started. The investigation delved into the C hook that failed, the maintenance procedures at PG&E, and more. Among the facts unearthed: "Many components" on the transmission tower that provided the ignition spark dated back nearly one hundred years, to the original construction by the Great Western Power Company. An examination by the California Public Utilities Commission found that PG&E failed to sufficiently inspect and maintain the aging transmission lines—a dangerous omission, considering their proximity to a fire-risk area.[47]

The result was a criminal indictment of Pacific Gas & Electric. In 2020, the company pleaded guilty to one count of unlawfully setting a fire and eighty-four counts of involuntary manslaughter.[48] Faced with billions of dollars in potential liability, the utility filed for Chapter 11 bankruptcy protection.[49] PG&E completed a restructuring and exited bankruptcy in July 2020—then one year later found itself in trouble again. That's when the Dixie Fire broke out, less than a dozen miles from where the Camp Fire began. The Dixie Fire claimed only one life but became the second-worst California fire ever in terms of acres burned. The cause, according to investigators: a tree falling on PG&E electric lines.[50]

As with so many disasters, the Camp Fire resulted from a number of problems that converged to produce a calamity. If only the C hook on the aged transmission tower hadn't failed. If only the location where the blaze began had been more accessible to the firefighters who arrived so quickly. But underneath it all, there's weather, an unavoidable factor. Drought conditions dried out the plants in the area, supercharging their potential as fuel. Wind spread the fire to nearby towns so quickly that people weren't able to escape. It might not be possible to completely eliminate the risk of wildfires, but knowing the weather, having intelligence on how dry things will be, where the winds will come from, and how powerfully they will blow, is crucial in reducing the danger.

"It's getting worse," says Brian D'Agostino, the San Diego Gas & Electric weather czar, as he shows me some weather maps in a glassed-in conference room looking out over the emergency operations center. "The fires are getting more destructive and becoming a bigger threat to life and property." That might not be a controversial statement, given the recent history of California wildfires and the link to warmer and drier conditions from climate change, but coming from D'Agostino, it's a sobering one. He has been immersed for more than a decade in the weather of Southern California and knows its potential to cause havoc. Growing up in the suburbs of Boston, he was a self-described weather geek, at first more fascinated by East Coast nor'easters than the Santa Ana winds. After college he became a broadcast meteorologist and found work in Montana, where the wildfire-prone landscape nudged him deeper into fire weather. In 2009, he took his forecasting skills to SDG&E.

It was a critical moment for the utility company. In 2007, SDG&E had experienced a serious wake-up call. That was the year the Witch Creek Fire devastated San Diego County. It began on October 21 in Witch Creek Canyon, near Santa Ysabel, a community less than forty miles from downtown San Diego. From the start, the Santa Ana winds pushed the flames westward to more populated areas. The next day, the Witch Creek Fire merged with another blaze, the Guejito Fire. And days later, it joined the Poomacha Fire.

Fatalities were limited to two people, but the impact on the San Diego region was astounding nonetheless. County officials later assessed that more than 197,000 acres, or over 300 square miles, had burned. Some 9,250 acres were within the city of San Diego itself. Over 1,100 residences were razed, including 365 inside city limits.[51] The Witch Creek Fire drove the largest evacuation in the county's history, with over half a million people told to leave their homes. And the cause? Utility lines were identified by investigators as ignition sources for both the Witch Creek and Guejito Fires.[52] In the aftermath, SDG&E moved to expand its in-house meteorological capabilities. "That's where there was a corporate decision made," D'Agostino says. "No matter what happens, this will never happen again."

D'Agostino continues: "We built our own weather infrastructure. To state it succinctly, we built the largest utility weather network that existed anywhere in the world." One front of that effort meant gathering observations. SDG&E has deployed 220 weather stations throughout the region, forming the company's own local sensor network and giving it a highly granular picture of conditions. As D'Agostino explains, because SDG&E has set up this network itself, the company has been able to site its weather stations where they're most relevant to the utility's operations and power lines. D'Agostino and his colleagues get data from the weather sensors on temperature, humidity, wind speed, and wind direction, and in addition, they can actually see what's happening on the ground. Some 120 cameras dot the hills and valleys, providing live feeds to help them spot trouble, such as signs of smoke.

In the conference room behind the emergency operations center's rows of workstations, D'Agostino calls up a feed to demonstrate. This view is from one of SDG&E's newest and most advanced cameras. Perched near the top of Cuyamaca Peak—the second-highest point in San Diego County—the camera shows a high-definition image of the mountains and valleys from an elevation of 6,493 feet. D'Agostino can pan and zoom to look more closely at potential problem areas. The view underscores how complicated the weather here can be. "Right here, we're higher than the top of Mount Washington," D'Agostino says, referring to the New Hampshire peak popular with tourists and hikers that is the highest spot in the northeastern United States. "And we're in San Diego County, we're in our SDG&E service territory. I think a lot of folks don't fully appreciate the scale of this topography. If you zoom around, you're looking down into totally different climates, deserts, slopes, foothills, and communities."

If SDG&E has gotten sophisticated about how it watches the weather, it's even more advanced when it comes to predicting it. The company feeds the data from all those weather stations into its own in-house forecast models running on supercomputers. Just as with the nationwide numerical weather-prediction models operated by NOAA and the National Weather Service, these models take data describing current conditions and crunch the numbers to simulate the weather moving forward. On top of that, SDG&E uses artificial intelligence techniques that draw on its repository of past weather conditions to tweak the modeling. AI requires large sets of reliable data to train the algorithms, so this archive is an asset. "The benefit of starting ten years ago is now we have an enormous amount of data, right? We've got two hundred and twenty weather stations reporting for a decade on every ridge and valley," D'Agostino says.

The model projects ninety-six hours into the future. "That gives us pretty much four days of peak-winds forecasts at all of our weather stations," D'Agostino says.

Winds are necessarily the focus of any effort to predict fire risks. Wind can further desiccate the vegetation in warm and dry conditions, supercharging the plant matter's fuel potential. Wind can feed oxygen to a nascent fire, ensuring its growth. It can also push the fire, potentially moving it in the direction of additional fuel and carrying away embers that light up additional blazes. Wind also plays a part in how fires get started, with particular implications for power utilities like SDG&E. Strong gusts can bring down power lines or blow tree branches and other vegetation into otherwise stable wires. Either way, sparks can result.

To underscore a basic effect of wind—the ability to supply oxygen—one needn't look further than what happens when you blow on some kindling to get a campfire going. The Greeks considered fire one of the four basic components of the physical world, along with earth, water, and air. Thanks to chemistry and physics, our understanding of fire has advanced since then. We know that fire is an exothermic chemical reaction—that is, one that releases energy in the form of heat. Fire requires both a fuel and an oxidant, which, outside of a lab, is usually the oxygen in air. In burning, the fuel and oxidant will react in a chemical change. A simple example: Burn hydrogen gas (H_2) as fuel in the presence of oxygen (O_2) and you get heat plus water (H_2O). Burning more complex substances produces other chemicals and by-products, resulting in smoke. But one way or another, you need a source of oxygen.

San Diego County and the rest of Southern California are especially vulnerable to fires because of the Santa Ana winds. The Santa Anas, which are most prominent in autumn, are believed to have taken their name in the nineteenth century from the Santa Ana Canyon south of Los Angeles. In meteorology lingo, the Santa Anas are *katabatic* winds—which simply means they result from air flowing downhill. (The term *katabatic* is derived from a Greek word meaning "descending.") When high pressure forms over the Great Basin region to the north and east—principally Nevada—cool air sinks. The movement of air around the high-pressure area brings winds that travel from east to west, carrying cool, dry air from the desert across the mountains and toward the Pacific.

As the air travels down the slopes, it gets compressed, which heats it up. The air picks up speed as it gets channeled through canyons. The result: hot, dry winds of exceptional force that go on for hours and sometimes days. Santa Ana gusts can routinely top forty miles per hour and have been measured at over one hundred miles per hour in some instances. On February 26, 2020, the National Weather Service recorded a 106-miles-per-hour gust at Sill Hill in the mountains northeast of San Diego.[53] If winds at those speeds were sustained rather than intermittent, they'd be the equivalent of a hurricane.

When the weather team at SDG&E looks over the output from their computer models, they're paying special attention to the wind—how much, where, and how long it will last. What can they do about it? I asked D'Agostino to walk me through what happens at the utility when the forecast indicates risky weather ahead. At a week or more in advance of predicted fire-danger conditions, it's mostly a matter of readiness. A big Santa Ana wind event calls for all hands on deck, so D'Agostino and his colleagues will review schedules and get ready to alert key personnel throughout the company's operations. If the forecast continues to warn of fire weather, SDG&E can take a number of steps, such as inspecting equipment in high-risk areas and checking to make sure vegetation isn't too close to poles supporting power lines. Inspections can involve helicopters and drones as well as staff on the ground.

But the big tool, one that residents in fire country dread, is the public-safety power shutoff, or PSPS. The logic is simple: If there's no electricity flowing through lines, they can't spark and cause fires. When faced with particularly high-risk conditions, utilities can turn off the power in a PSPS. The trade-offs are difficult. On the downside, customers, from individual homes to businesses, health-care facilities, schools, and more, may wind up in an intentionally ordered blackout, with all of the problems that suggests. The upside is avoiding a catastrophe like the Camp Fire. But if utilities lean too hard on power shutoffs, customers will complain, and regulators will scrutinize the decisions.

Weather forecasting can do more than simply call out fire-prone conditions in advance. A good forecast can home in on the danger zone, helping utility managers limit the duration and number of customers covered by a PSPS. D'Agostino calls up a map to show me the SDG&E forecast model's prediction for winds the next day. A color-coded overlay displays the forecast for wind-gust speeds. In another layer on top of that, tiny animated curving arrows depict the wind direction and movement. The arrows flow endlessly, tracing out paths around mountains and through valleys. The effect is mesmerizing. You can see this for yourself online at the website for the San Diego Supercomputer Center, wxmap.sdsc.edu. Other SDG&E weather forecasting can be found at sdgeweather.com, and feeds from those observation cameras are available to all at alertcalifornia.org.

The reality around high-stakes decisions such as shutting off power is more complicated than a single forecast. For example, D'Agostino's staff will consult other sources, such as the Santa Ana Wildfire Threat Index (SAWTI). The SAWTI (available at https://fsapps.nwcg.gov/psp/sawti) factors in not only weather but also how dry the fuel—that is, the plants, bushes, and trees—will be. SAWTI was developed by the US Forest Service, the University of California, Los Angeles, and SDG&E. Such collaborations are important. Indeed, the map of Southern California is dotted with an entire ecosystem of organizations involved in combating the wildfire threat. To observe one of them, I took a short drive up Interstate 15.

I am looking out a broad wall of windows, enjoying a spectacular sunny view. The National Weather Service's San Diego weather forecast office operates out of a suburban office park north of downtown, sharing the complex with a real estate company, a law firm, and some financial planners. The vista out these windows easily makes up for the lack of downtown glamour. From the office's hillside perch, the meteorologists can gaze inland across subdivisions and businesses in the valley below all the way to Palomar Mountain in the distance. The panorama encapsulates the kind of terrain that makes fire weather one of the top preoccupations at this particular outpost of the Weather Service.

"San Diego is not the place you go for weather, for *real* weather," says Alex Tardy, the warning coordination meteorologist here. The picture-perfect day outside proves his point. Tardy himself was raised on a diet of

Vermont winters, went to college at the respected meteorology program at the University at Albany, then did stints watching snowstorms in Salt Lake City, Utah, and hurricanes and flooding in Corpus Christi, Texas. "Fire is our number one problem. But we have people who don't think that fire weather is *weather*," Tardy says. "Even people who watch TV and see large fires in California, they don't necessarily think that's weather-related. They don't comprehend what *low humidity* means. They don't comprehend what happens when you have low humidity and wind on a fire. Or when you have a drought or a normal dry summer, what that does to vegetation. They know what it's like to be thirsty, but they don't understand what it's like for vegetation to be thirsty. I don't think they can comprehend that."

Tardy knows better. He uses an example from his time in Texas for a comparison. Hurricanes draw a massive response. The playbook is familiar: TV forecasters point to track maps days in advance, the public is advised to stock up on food, water, and medications, then there is news of evacuations, rescues, and cleanup. "But you take a wildfire situation where fires are occurring and expected to get worse because of weather, and the amount of people involved here in California exceeds a hurricane response," he tells me. "Because in a hurricane, much of the time there's nothing you can do about it. You have to let it come. You evacuate. But you're not fighting it."

If the work at SDG&E's emergency operations center emphasizes the power of preventive action by utilities, then the forecasting at Tardy's National Weather Service office serves as a reminder that the atmosphere is beyond human control—which means that all the people in the area need to know about risky conditions so they can use appropriate caution. Preventing blazes triggered by power lines is worth almost any effort, as demonstrated by the cost in lives and property from the Camp and Witch Creek Fires. But a fateful spark can come from anything—and it often comes from careless people. After a 2020 blaze in Solano County, California, southwest of Sacramento, that spread out over three hundred acres, fire officials found a cigarette butt they concluded was the cause.[54] A loose chain dragging behind a truck, a drooping muffler scraping the pavement, or a tire blowout that causes the metal wheel to grind on the road can throw off sparks, so warning drivers to check their vehicles has become an annual ritual in at-risk areas. And of course, there's that lost hunter whose signal fire unleashed the Cedar Fire.

The most obvious public-facing wildfire work of the San Diego WFO takes place when forecasters issue a fire-weather watch or a red-flag warning. The watches and warnings reflect a combination of windy and dry weather that ups the risk. The San Diego office will issue a watch if red-flag conditions are expected over the next one to three days. The actual warning comes when sustained winds will exceed twenty-five miles an hour, gust above thirty-five, or both, while relative humidity drops below 15 percent.[55] For all of the scrutinizing over forecasts behind the scenes at SDG&E, the red-flag warnings represent the critical official signal to the public—particularly those residents who may be irate when they learn that their electricity will be out due to a public-safety power shutoff.

But beyond the public warnings, there's a lot happening behind the scenes at the Weather Service. The fight against wildfires shows how the role of the NWS forecaster has evolved. Not that long ago, preparing the forecast during routine weather and deciding watches and warnings for dangerous weather made up the bulk of the job. But the accuracy of computer models has improved, which allows meteorologists to spend more time ensuring that forecasts get to the right people and helping them understand the implications of what's in the forecast.

"It used to be that the only messages we sent out were things like the red-flag product, and that was one hundred percent public, right? And then we would just be done," Tardy says. "We would just move on or monitor the weather. Now I would say there's probably fifty percent of our time working on messaging either through video, email, or phone calls leading up to events—that's fifty percent of our time on internal messaging." By *internal*, Tardy doesn't mean within the Weather Service; he means the entire constellation of agencies, organizations, and companies in the emergency-management community for the San Diego region. That spans everything from the Weather Service to the US Forest Service to, yes, San Diego Gas & Electric. And, unavoidably, it extends to the people who have to risk their lives fighting the fires once they've started.

"In Southern California, or all of California, if you're a firefighter or a chief or anybody that has to do any kind of incident response, then you're a weather geek," says Cal Fire's Suzann Leininger. "We all have to pay attention to weather." Leininger's official title is research data specialist, which is a fairly flat way of describing what her job actually entails: sifting through computer models and real-time information to figure out how a wildfire will behave, the better to stop it in its tracks.

Meeting with Leininger, I find myself in yet another command center, a semi-darkened room with workstations and big screens all around. We're at the San Diego unit headquarters for Cal Fire, a group of low buildings perched well east of the city, just before the suburban sprawl gives way to the mountains. One workstation includes six different computer displays as well as microphones on flexible gooseneck stands, a reminder that during an emergency, the people working here will be communicating with firefighters in the field. Today this auxiliary room is quiet but in the main communication center next door, staffers of Cal Fire and the US Forest Service are going about their work.

Cal Fire traces its origins back to the nineteenth century. To understand how, it's useful to remember the organization's full name: California Department of Forestry and Fire Protection. In the 1800s, the opening of the West spurred concerns about both conservation of wild areas and management of forests, given the value of lumber to economic growth. Both factors drove federal and state forestry efforts. In 1885, the California legislature set up a forestry board. Then in 1905, lawmakers created the position of state forester and gave the person who held that job the power to appoint fire wardens, according to an in-house Cal Fire history.[56] By 1929, the state forestry agency was buying its first fire trucks. And in the 1940s, under Governor Earl Warren, the forestry agency was explicitly charged with providing fire-protection services on state-owned lands and authorized to deliver those same services to counties if they so desired (and agreed to pay). As the in-house history puts it: "The California State Government was now inextricably in the business of wildland fire control."

Today Cal Fire watches over state lands as well as many local areas through "cooperative fire protection agreements" with city and county governments, including San Diego County. To supplement the agency's personnel in major incidents, it draws on prison inmates in the Conservation Camps Program with the state's Department of Corrections and Rehabilitation.[57] The Cal Fire fleet includes over three thousand fire engines and other types of equipment, from bulldozers to mobile kitchens that can each feed up to five thousand firefighters and evacuees a day. It flies more than sixty airplanes and helicopters in what Cal Fire describes as "the largest civil aerial firefighting fleet in the world."

Figuring out the best way to deploy all those resources is critical, which is where Leininger and her colleagues come in. Leininger may be a weather geek, but she's not a meteorologist. Her job cuts across multiple disciplines. She's the fire world's version of an intelligence analyst.[58] Leininger is a volunteer firefighter herself, so she's well positioned to make sure firefighters get the information they need and don't get flooded with what they don't.

One of Leininger's tools is a powerful computer system called Wildfire Analyst. The software predicts how a wildfire will spread, drawing on current weather conditions, forecasts, and information about terrain, roads, vegetation, and moisture. Those few words—*terrain, roads, vegetation,* and *moisture*—don't do justice to the staggering work that goes into collecting such data. For example, mapping out the location and density of plants and trees involves analyzing satellite imagery as well as aerial laser scans that create three-dimensional maps.

Wildfire Analyst is the creation of Technosylva, a company based in San Diego and León, Spain, dedicated entirely to wildfire preparation and response. Technosylva works closely with practically everyone in the Southern California wildfire ecosystem: Cal Fire, SDG&E, other utilities, and the Forest Service. The company's founder and chief technology officer, Joaquin Ramirez Cisneros, joined me and Leininger in the emergency command center to talk about where weather fits in wildfire prediction.

"The biggest drivers are weather and climatology, and those are two different things," says Ramirez, who studied forestry in Spain before starting Technosylva and remains on the faculty of Universidad de León. Forecasters must account for both the near-term potential for wind and the weeks and months of normal or drought conditions that determine whether shrubs have dried to tinder. "We need to know how is our fire season compared to past fire seasons. We need to know what happened in the last three days. We need to know what's going to happen in the next six days."

Progress, Ramirez says, comes from modeling the weather at ever-finer scales of distance and time. The staple high-resolution weather models feeding into Leininger's computer have a resolution of two kilometers and show expected conditions in one-hour increments out to several days. "Two kilometers is such a long distance," Ramirez says. "You have canyons right here in San Diego that are two kilometers long but they may have a lot of terrain in the middle. The reality is that those weather models that the meteorologists use, they're not even what we need. We need to do more." Technosylva takes the information from those two-kilometer models and uses mathematical techniques to downscale the data to two hundred meters to predict the small-scale eddy currents of winds shaped by the terrain. Wildfire Analyst uses all of this information to simulate how a blaze will spread, informing decisions about how potentially dangerous a given fire will be and how and where to fight it.

Both Leininger and Ramirez emphasize that, crucial as the minute details of the weather forecast may be, they need to stay here in the command center. Firefighting teams in the field need bottom-line, actionable advice. Ramirez pulls up an iPhone app for Cal Fire that ties into Wildfire Analyst. It shows a readout for a simulated fire, headed by a bright color-coded "Initial Attack Assessment" that "evaluates the difficulty of the incident considering the first hour of spread" on a one-to-five scale. "We always say, big gloves, big buttons," Leininger says. "They're wearing gloves and they need to be able to operate the app." Ramirez calls it "fat-finger technology."

Captain Neil Czapinski, a veteran Cal Fire firefighter, describes the perspective from the field: "All the intel is great, and more intel is even better. But at a certain point, you've got to quit looking at the intel and start fighting the fire." Czapinski adds that for all the progress in detecting risky weather in advance to prevent fires from igniting and all the advances in modeling how weather will cause a fire to grow, one aspect defies prediction. "It's that human factor, the emotional attachment that makes people do what they're going to do," Czapinski says. Sometimes residents ignore recommendations to evacuate, with tragic results. "With all the warnings in the world, there's still going to be people lingering."

Better weather prediction unquestionably helps prevent wildfires. It helps combat the fires that do occur. But even as these solutions move forward, the problems keep expanding too. Climate change brings warmer and drier weather to the fire-prone areas of the western United States, upping the risks. And unchecked development puts more people in harm's way. Those two factors combined will drive a predicted increase in extreme fires worldwide as high as 14 percent by 2030 and over 30 percent by 2050, according to 2022 projections by the United Nations Environment Programme and GRID-Arendal, a Norwegian environmental nonprofit.[59] The authors of that report called on governments and organizations worldwide to emphasize preparation and prevention over fighting fires. "Wildfire risk reduction activities represent a sound return on investment as they reduce the potential impacts of wildfires," the report said. "In the long term, they will be more cost effective than relying on reactive firefighting and post-disaster recovery efforts." In that sense, the integration of cutting-edge meteorology with other fire strategies in San Diego County offers a model worthy of attention.

In California, officials warn that extreme fires—defined as those that burn more than 10,000 hectares, or 24,710 acres—could occur 50 percent more frequently by the end of this century as compared with the period from 1961 to 1990 if "high-emission" climate scenarios come to pass.[60] In other words, failing to curb carbon emissions will translate to more of the worst fires.

Forecasts that provide warnings of potential fire weather a week or more in advance are helping utilities like SDG&E and others that are increasingly building their in-house meteorology capabilities to prepare effectively. Facing dry, windy weather—especially winds like the Santa Anas in Southern California and the Diablo winds in Northern California—power companies can put staff on notice, step up equipment inspections, and implement power shutoffs when necessary. If power shutoffs must occur, using better forecasts to narrow down where they're needed and how long they must last can reduce the number of those affected.

But in the long term, with climate and development risk factors growing, it's difficult to view pulling the plug and leaving homes and businesses without electricity as a viable solution. It's forcing utilities to invest in massive infrastructure changes, such as burying power lines instead of suspending them over flammable shrubs and trees. In 2021, Pacific Gas & Electric announced a ten-year plan to move ten thousand miles of its lines underground at a cost of billions. Another infrastructure approach is the creation of microgrids, sections of the power grid that can operate independently using battery storage or local generators to supply key locations, such as schools, hospitals, and shelters, with electricity when the larger surrounding portion of the grid has been turned off. And as more individuals adopt clean-energy solutions, they will have options to provide their own electricity through solar setups or even by using electric cars as backup batteries to power their homes.

When the inevitable fires do occur, progress in short-term forecasting can help contain them more effectively. A research project at the National Center for Atmospheric Research (NCAR) in Boulder, Colorado, promises computer modeling of the atmosphere at levels of detail even finer than the kind of high-resolution systems being used at Cal Fire—down to a level of five meters, or just over sixteen feet. To accomplish this, the NCAR project leverages the same technologies that power video games and other high-end graphics. In all computer modeling of the weather, there's an inherent trade-off in how the software simulates the atmosphere. If you want to zoom in more closely to see how the weather will behave in small areas, you need more computing power. If you want to look further into the future, you need more computing power. To predict conditions days in advance, the big global-forecast models break the atmosphere up into areas measured in kilometers in order to simulate the weather across the entire world. Higher-resolution models like those used by SDG&E can focus in tighter.

FastEddy is a computer model that makes it possible to zoom in even more by tapping the power of special computer chips known as graphics processing units, or GPUs. Computers large and small (including your own desktop or laptop) typically use central processing units, or CPUs, as their main brains. As graphics have become ever more complex and realistic—and taxing on computer processors—makers developed a way to take some of the load off the CPU with separate chips optimized to handle those graphics: the GPU. As GPUs evolved, engineers came to realize they could help provide the capabilities of traditional supercomputers at a much lower cost.[61]

"The key to the type of modeling we do is that we want fine temporal and spatial scales," says Jeremy Sauer, an NCAR scientist who leads the

FastEddy project with his colleague Domingo Muñoz-Esparza. By modeling the behavior of air in very small units, FastEddy can account for things like the effects of terrain and tree canopy. "All of these things are typically well below the spatial scales that are captured by numerical weather-prediction models," Sauer tells me. "Why? Because computing was never significant enough to do it. But with these emerging hardware technologies, we can actually achieve such tremendous speed that now we can actually talk about forecasting at what we call the microscale."

Translation: The global-weather models, like those used to predict how hurricanes will evolve days in advance and whether a heat wave is on the way, can't tell you how the wind will behave at, say, two different ends of a small canyon. And the global models don't need to, because those small phenomena don't affect large-scale movements in the atmosphere enough to change the big picture. Think of it this way: At the beach, a boulder on the shore's edge will shape the way a wave rolls over it and affect the water in the immediate vicinity. But a hundred yards out, the boulder doesn't matter; the waves keep on rolling in.

Sauer brings to this research a formidable mix of computer and weather knowledge. He earned an undergraduate degree in physics and computer science at the University of Montana, then followed up with a master's degree in computer science and a Ph.D. in geophysical fluid dynamics. He also worked early in his career at the Missoula Fire Sciences Laboratory in Montana, a key research center of the US Forest Service. That helped stoke his interest in microscale forecasting. "I was working on wildland fire behavior, and modeling that, in a big way, is influenced by the way the wind blows close to the ground," he recalls. "That's where I started to get into these notions of small timescales and small spatial scales, because they're so important to fire behavior." That fine resolution can capture the winds whipping around hills, the better to predict how a wildfire would spread.[62]

My last stop in San Diego provided a reminder of the human stakes underlying all this work. In another fairly unremarkable office-park building, I sat down with Ray Chaney, a retired Cal Fire chief who has been on the business end of countless 911 calls. His experience includes overseeing an air attack base and acting as an airborne spotter, observing fires from above to help direct the response. Now he's the director of emergency services for 211 San Diego, the nonprofit organization that serves as a clearinghouse for community services in the region.

If you've never needed to use a 211 service, you may not appreciate its role. Most cities and regions in the United States have some version of it. Many are operated by independent nonprofits, such as United Way. (Where I live, in New York City, the analogous service is 311, and it's run by the city government.) If you need help in an emergency, you call 911. If you need help in a nonemergency situation, you can call 211. At 211 San Diego, those calls cover a distressing range of the problems people can find themselves confronting. Operators provide referrals to drug-treatment programs, help people get food benefits, and locate shelters for homeless individuals. The service's operators are trained to understand these situations so the callers get the assistance they need.

Those are what might be considered the routine calls, though it feels unseemly to label them as such. To the individuals facing a crisis, they are anything but routine. Nonetheless, there's a contrast between those everyday inquiries and what happens in a larger-scale incident such as an earthquake, flood, or fire. Chaney explains that, faced with a possible disaster, 211 San Diego begins ramping up to respond. As with almost any big-city emergency community, all the key players know one another. So when a red-flag warning looms, 211 San Diego starts its own planning in parallel with steps being taken elsewhere in the wildfire weather ecosystem. Chaney will get forecast briefings from Alex Tardy and his team at the National Weather Service and he'll hear about the likelihood for power shutoffs from Brian D'Agostino and his colleagues at SDG&E. At 211 San Diego, Chaney can alert staff and volunteers about the potential need to have extra people answering calls in the days to come. They can change the phone tree of the interactive voice-response system that funnels the incoming calls. (As in *If you're calling about a public-safety power shutoff, press 1.*)

"Our call center does an intake on them. Then we do a risk-rating scale and figure out what their needs are," Chaney says. "And then we can do a referral to get them a generator. Or get them transportation out of the region. We can let them know where they can find a community resource center near their area where they can get food and water supplies to help during the shutoff. Or a hotel voucher to say 'Look, let's just get you and your family out of the area for a couple of days until power can be restored. We'll get you transportation there, and we'll get you back once the incident or the weather event has been mitigated.'"

The distance from arcane computer models and wind predictions to lives either helped or disrupted is not so far. Better weather forecasting helps lower the chances of worst-case events like the Camp Fire that result in people dying in their homes and cars. But the example of San Diego County demonstrates the effort involved. Meteorology must become embedded throughout the organizations responsible for stopping fires, and big companies must, like SDG&E, stand up their own forecasting centers. And no amount of progress on the weather-prediction front should obscure the extent to which climate change threatens to undercut these efforts. The preventive measures aren't trivial. Power shutoffs are an inconvenience for some but a medical emergency for those who depend on devices or refrigerated medicines. Evacuations can be an annoyance for some but an economic upheaval to those without a financial cushion. Success with short-term mitigation will need to be paired with long-term work spanning carbon reduction, sophisticated forestry, and more sensible approaches to development. In California, no matter how sunny the weather might seem, there's always smoke on the horizon.

3 The Local Forecast

INSIDE YOUR EVERYDAY WEATHER REPORT

"Sending the AFD!" It's 10:56 on a clear summer night at the State College, Pennsylvania, office of the National Weather Service, and that terse and somewhat cryptic announcement encapsulates some of the most important work that takes place every day, around the clock, here and at Weather Service locations across the country. AFD stands for "area forecast discussion." John Banghoff, a young meteorologist on duty, has just announced to his colleagues that he has completed tonight's AFD and is sending it out to the world. From the AFD, anyone who cares about the weather in central Pennsylvania will learn that tomorrow should bring "highs ranging from the mid 80s over the highest terrain of the Alleghenies, to the low and mid 90s in the Susq[uehanna] Valley." The start of the weekend promises heat, humidity, and "ample sunshine." Pretty good news for anyone planning on taking the kids to the roller coasters at Hersheypark or hiking Mount Nittany on Saturday, even if it means keeping hydrated and bringing along the sunscreen.

We live on a big planet. Predicting the weather requires a global view because the Earth's atmosphere is a global system. NOAA and the Weather Service operate a number of important facilities that track the whole world and forecast what's coming to the United States, which I describe in more detail in other parts of this book. There's the Storm Prediction Center in Norman, Oklahoma, that watches the country for severe thunderstorms and the convective weather that can result in tornadoes. In College Park, Maryland, the Weather Prediction Center monitors the potential for precipitation and follows the pressure systems that will shape the weather in days to come. NOAA's Environmental Modeling Center makes sure the computer-forecasting systems churn out the data-driven projections of weather across the globe.

But for *your* weather—the forecast that helps you decide whether to wash your car or take an umbrella when you head out the door—the place that matters most is your local weather forecast office. It's where all that other information funnels in to inform local meteorologists with a local mission. Even if you get your forecasts from your local TV or radio station, the Weather Service outlook often forms the foundation for those pronouncements. WFOs represent the critical link to communities at the end of the forecasting chain.

An AFD gets produced and published at least twice a day here at State College and at every one of the other 121 weather forecast offices of the National Weather Service. Forecasts for my home in New York City, for example, come from the New York WFO on Long Island. Across the Hudson River is New Jersey, and most of its weather gets handled by the Philadelphia office—which is actually located in Mount Holly, New Jersey. Depending on where they live, residents of my native state of Michigan get information from one of five different WFOs: Detroit (technically, the office is located in White Lake, near Pontiac), Gaylord (for the northern part of Michigan's mitten plus a slice of the Upper Peninsula), Marquette (for most of the Upper Peninsula), Grand Rapids, and, for a sliver of southern Michigan at the state border, the northern Indiana WFO. Hawaii has its own WFO, as does Guam; Alaska has three.[63]

The State College WFO, where I'm watching John Banghoff issue the forecast, covers the mountainous center of Pennsylvania, roughly from Johnstown all the way east to Lancaster County, before handing off coverage to the Philadelphia WFO. I'm here with Banghoff today to get an inside look at the core work of the National Weather Service. When any type of weather danger threatens, the work of a WFO takes place in a very different register. Then, these forecasters form the front lines of emergency communications, tasked with the responsibility of deciding whether to issue official warnings about urgent hazards like tornadoes or flash floods. On days when those storm clouds gather, Banghoff and his colleagues must monitor their screens, track weather movements minute by minute, and let the public and local officials know about any dangers. "Severe weather events, those are probably our busiest, most active days," Banghoff says. "You can be issuing warnings for eight hours in a row."

Today is not one of those days. This is a day with what you might call normal weather, if such a thing can be said to exist. There are no emergencies on the horizon. Banghoff and his colleagues will go about their work, looking over computer maps and scanning radar displays. They'll apply their meteorological training and their experience and knowledge about the way weather behaves in central Pennsylvania to issue forecasts. And they'll attend to a hundred other unseen tasks that a weather forecast office takes care of every day to support the people and organizations in their coverage area, including some you've probably never thought about.

Banghoff's eight-hour shift doesn't begin until 3:30 p.m. But after a quick lunch, he kicks off the day a few hours early by joining up with two colleagues at the office and heading over to Beaver Stadium, the beloved temple of Pennsylvania State University football. Banghoff and his NWS colleagues are meeting with officials from the university's public-safety and emergency management team. I follow them through a door marked POLICE into a warren of offices. The group convenes in a large but windowless room that normally serves as a cafeteria but would be pressed into service as an emergency operations center, complete with phone hookups and videoconferencing, if needed.

The topic for this group: weather readiness at Penn State. There are some 48,000 students at the main University Park campus in State College, and the school has twenty-three other campuses across Pennsylvania. In fall of 2023, the entire Penn State student body numbered 87,903. The university says it has over 36,000 full- and part-time employees. Put it all together, and Penn State public-safety officials have a responsibility for a population equivalent to a midsize US city, more than Wilmington, Delaware, or Green Bay, Wisconsin.

First on the agenda is a discussion about maintaining Penn State's certification as a StormReady institution. The idea for the StormReady program dates back to 1998, when Steve Piltz, a forecaster in the Tulsa

WFO, wanted a way to motivate local emergency managers to increase their weather preparedness. For emergency managers, the certification would elevate the visibility of weather dangers inside their organizations and also offer a chance to earn some personal recognition by meeting a set of standards. For the Weather Service, the program would help WFOs with their mission by publicizing best practices for safety, particularly among those in a position to implement them.

According to a brief history of StormReady by Lans P. Rothfusz, one of the program's developers, the originators sought to dispel the idea that forecasts on their own could prevent disasters.[64] "Frequently, the NWS is viewed as the problem when a community is caught by a 'surprise' severe weather event," Rothfusz wrote. "While the inexactness of the science sometimes contributes to these 'surprise' events, all too often the 'surprise' is the result of multiple failures within the entire warning system: NWS, emergency managers, the community, and the public. Some of these failures are the result of inadequate preparation before the event occurs." Over two decades later, more than 1,600 counties, over 1,200 local governments, and hundreds of universities, manufacturing sites, and other organizations have been designated StormReady. Piltz and Rothfusz went on to be awarded the Commerce Department's Silver Medal for creating a national standard of weather preparedness.

To receive StormReady certification, organizations must have, among other things, a twenty-four-hour emergency operations or communication center and multiple ways to receive weather warnings and disseminate them within their communities; they must conduct training and hold exercises.[65] Like most big universities, Penn State has text, phone, and email systems to alert students and staff about emergencies. Likewise, the public-safety officials conduct regular exercises related to weather emergencies. The Weather Service team assures the university officials that they don't foresee any obstacles in recertification.

Game days are the other big topic. Beaver Stadium is quiet on this early summer afternoon, but when Penn State plays at home in the fall, more than a hundred thousand fans fill the stands. Game days bring all the trappings of college-football fandom: Alumni traveling from hundreds of miles away, parking lots jammed with RVs and cars, and elaborate tailgate parties with tents, buffet tables, games of cornhole, and, of course, beer.

Blue-and-white flags fly everywhere, most displaying the stylized image of the Penn State mascot, the Nittany Lion. (The lion logo depicts an eastern mountain lion; Nittany is taken from nearby Mount Nittany.) Game-day traffic is infamous; the public-safety team arrives at six a.m. and expects to work a sixteen-hour day.

It's all good until the skies darken in Happy Valley, as the region encompassing State College is known. Whether the day may bring heavy rain, high winds, or snow, university officials must monitor conditions closely in case of severe weather that requires a stadium evacuation. Thunderstorms with lightning represent a particular danger with one hundred thousand fans seated out in the open. For any big venue, evacuations are enormous undertakings. Unnecessary evacuations are inconvenient at best, and they carry their own risks of people falling or getting injured if panic makes the process turn disorderly. Here at Beaver Stadium, an evacuation would send some fans to nearby enclosed structures, such as the Bryce Jordan Center arena, where men's and women's basketball games are played. Those in covered areas of the stadium could shelter in place.

To keep everyone safe on game days and maintain awareness about changing weather that could affect fans and players, Penn State contracts with AccuWeather, the privately held forecasting giant headquartered a few miles away from the University Park campus. University officials spoke well of the forecasting they got from AccuWeather. They told me about one game when storms had formed relatively close by but the AccuWeather meteorologist advised them there was no threat to the stadium area. Based on that confidence, the public-address announcer reassured fans that any lightning they might see in the distance posed no danger.

The Penn State officials recount that anecdote to emphasize the value of what they termed a "good call"; knowing when it's safe to stay put carries its own worth. But with so many people at the games, officials are looking for belt-and-suspenders redundancy. And when it comes to launching an actual evacuation, they like having the weight of an official forecast from the National Weather Service to back up the move. The Penn State and NWS representatives agree on a plan to have a meteorologist from the WFO assigned to monitor each game. To the NWS team, there's no question that game-day weather is a matter of public safety, a key part of the Weather Service mission. There are just too many people in one place outdoors in a small area. Banghoff puts it this way: "A home football game in Beaver Stadium is perhaps the biggest weather vulnerability in our entire area all year long."

That decision and the meeting itself illustrate the evolving role of the National Weather Service. For decades, the staffers at a WFO focused practically all their efforts on examining weather data and issuing forecasts, often in a *Dragnet*-style "just the facts" approach. But time and again, after-action reports on major weather incidents have shown that good forecasting isn't enough. In 2011, NOAA launched the Weather-Ready Nation initiative, which includes an emphasis on decision-support services and community partnerships to improve preparedness.[66] Efforts like StormReady and Weather-Ready Nation stress the role of the WFO in working with local officials. It's a side of the meteorologists' work that takes place in every WFO, even if it is mostly unappreciated by the public.

Business concluded, the Penn State officials take us up an elevator for a quick tour of the Beaver Stadium press box. High above the field, we can view the entire bowl of the stadium. Today the bleacher-style seats are empty, but it's easy to imagine the noise of a capacity crowd cheering for the Nittany Lions on an autumn Saturday—perhaps one that's crisp and sunny, or maybe one where dark clouds drift in, bringing the dangers of lightning and wind to the fans below. As I picture every one of those seats filled, the burden of deciding whether to tell everyone they need to suddenly get up and leave feels daunting.

By 3:15 p.m., we're back at the weather forecast office, where John Banghoff gets ready to kick off his 3:30 p.m. to 11:30 p.m. shift. Sadly, there wasn't enough time after the Beaver Stadium meeting to grab ice cream at Penn State's legendary Creamery. (Using milk from Penn State cows, the Creamery has been producing dairy treats for more than one hundred fifty years, true to the university's roots as a land-grant college with a robust agriculture program.)

Banghoff begins by logging into various systems across five different computer screens, including the Advanced Weather Interactive Processing System, or AWIPS (pronounced "*ay*-whips"), the main workstation software for NWS forecasters. AWIPS gives him access to massive amounts of NOAA weather data, from satellite images and radar views to the National Blend of Models (NBM), the Global Forecast System (GFS), and other computer-model projections. It is the indispensable dashboard for any Weather Service meteorologist and can usually be seen displaying a dozen or more windows at once.

Weather forecast offices operate twenty-four hours a day, seven days a week, since conditions must be monitored at all times, and warnings may need to be issued at any moment. Here in State College, the rotation has three meteorologists on duty from 7:30 a.m. to 3:30 p.m. and another three from 3:30 p.m. to 11:30 p.m., the shift that's just getting underway. The overnight shift, from 11:30 p.m. to 7:30 a.m., is staffed by two. Banghoff explains to me the general division of labor for the three-member shifts. One meteorologist focuses on the short-term forecasts, loosely covering the next twenty-four to thirty-six hours. Another produces the medium- to long-term outlooks, forecasting weather four days or more in advance. To do this, the meteorologists draw on a broad range of information, from raw data output from the computer models and national-level guidance from the Weather Service's Weather Prediction Center and Storm Prediction Center to their own experiences with peculiarities of the area's climate.[67]

The third meteorologist, in what's referred to as the public-service shift, concentrates on communications, from fielding phone calls and disseminating weather updates to local officials to posting forecasts on social media. For this afternoon and evening, that's Banghoff. But the divisions are fluid, and the three forecasters occasionally discuss who's free to tackle tasks as the shift progresses. Forecasters rotate across the shifts so that no one works exclusively on nights or weekends. During periods of significant weather, additional staffers may be called in. Over an eight-week span, Banghoff can expect to work a stretch of seven overnight shifts in a row, as well as day and evening shifts.[68] Some forecasters have told me that constantly cycling through different shifts—and the attendant havoc on their sleep schedules—is one of the most difficult parts of the job, though it obviates the need for permanent overnight assignments.

Banghoff, dressed comfortably for his shift in blue chinos and a polo shirt with the Weather Service logo, is young, with short sandy-brown hair and a ready smile. He's energetic, part of the new generation of meteorologists, steeped in social media. He grew up near Columbus, Ohio, and attended Ohio State University, where he served as president of the meteorology club and graduated in 2015 with a bachelor's degree in atmospheric sciences and meteorology. He went on to earn a master's at Penn State in 2018 and joined the National Weather Service full-time that same year.

At one point Banghoff mentioned to me that he had been a member of the Ohio State University Marching Band, a renowned group known to Buckeyes as "the Best Damn Band in the Land." I later searched online in hopes of finding a clip of him playing the trumpet and discovered a video of a 2015 ceremony announcing the annual Most Inspirational Bandperson honor, voted on by the group's two-hundred-plus members: Banghoff. I wasn't surprised to see that. Banghoff is passionate about the weather, about his job, and about the responsibility of forecasters to help keep people safe. He's served as president of the local chapter of the American Meteorological Society, which is particularly active, given the weather nexus that is State College, home to not only a Weather Service WFO but also Penn State's enormously influential meteorology program and the headquarters of AccuWeather. He juggles all his duties with enthusiasm. I'm grateful to be able to watch Banghoff in action.

After firing up the computer windows he'll need, Banghoff digs into one of his first tasks for the afternoon: the Weather Story. This is a forecast summary for the public, written clearly and accompanied by lively graphics—a presentation meant for consumption on the web and social media. The format contrasts with what the Weather Service refers to broadly as *text products*. (In NWS lingo, a *product* is any type of forecast, alert, watch, warning, or other bulletin that is released to the public.)[69]

For a number of reasons, ranging from consistency and clarity to standards defined by the World Meteorological Organization, WFO text products get issued in a way that looks pretty old-fashioned in the smartphone era. Each forecast is topped by a complicated-looking set of codes that identify the type of message and its origin; for example, AWUS81 KCTP for a regional weather summary issued by the State College WFO, which is known by the three-letter code CTP. Text products often contain meteorological jargon. They might mention zonal flows, inversions, or advection. For professional meteorologists and even committed weather buffs, text products lay out the insights from the WFO in a clipped, no-nonsense tone.[70] They also serve as an archive over time.

The Weather Story aims to deliver something different. Whether it's for a member of the public or someone whose job requires awareness of potential weather issues, the Weather Story gives a quick but engaging and highly visual overview of what to expect. "There's creativity here," Banghoff tells me as he pulls up a set of PowerPoint templates used in the State College WFO office to build the Weather Story each day. "We're just trying to tell the story of what's going to be happening."

On this sunny July afternoon, the headline is heat. The State College forecasts call for highs in the low to mid 90s over the next few days and uncomfortable levels of humidity. That combination could produce values for the heat index—a measure used by the Weather Service that factors in humidity to indicate how temperatures affect humans—that flirt with levels that would trigger a heat advisory. Banghoff works with a series of PowerPoint slides that show color-coded maps of Pennsylvania displaying expected maximum heat index values across the state. There's a separate slide for each of the next few days. Next to the maps, he adds some specific recommendations: "Spend Some Time in Air Conditioning" (with an icon of a snowflake superimposed on a house) and "Avoid Strenuous Activity During Hottest Part of the Day" (with an icon of a stick figure hunched over some yard work). When he's done, Banghoff combines the slides into a GIF animation that rotates through them in sequence and uploads it to the CTP website. He titles the whole thing "Hot & Humid Week Ahead."

Apart from the use of graphics, the distinction between the traditional text products and formats like the Weather Story may seem subtle. But it underscores the evolution of WFOs beyond simply issuing the forecast to ensuring that information is understood and used productively—critical when it comes to doing something about the weather. Meteorologists like Banghoff increasingly concentrate on effective communication of forecast information, whether that's to specialized audiences like the Penn State public-safety team or to the public over social media. It has been one of the biggest changes at the Weather Service over the past twenty years and a priority moving forward. This work spans the needs of the entire region as well as specific groups with different vulnerabilities. As it turns out, I had earlier stumbled across one example without realizing it.

On one of my visits to State College, I left Interstate 80 behind to travel the central Pennsylvania back roads for a bit. Driving down a country lane, I passed a horse-drawn buggy. It's not an uncommon sight in Pennsylvania. The state is home to the largest Amish population in the United States. In 2023, nearly ninety thousand followers of the Amish religious tradition made their homes in Pennsylvania, according to the Young Center for Anabaptist and Pietist Studies at Elizabethtown College.[71] Dedicated to beliefs centered on humility and obedience, the Amish typically seek to keep modern society at a distance as a way to ensure their values are preserved. Part of that involves shunning technology, though the specifics vary and can be more complicated than most people appreciate. For instance, electricity supplied from the utility grid is generally prohibited, but battery power may be used in some cases. Televisions and radios are mostly off-limits.

I didn't give much thought to the Amish community's relationship with the weather until Banghoff conducted a routine test of the WFO's automated phone service—a real throwback in the internet age. When I was growing up, in the pre–World Wide Web era, calling a telephone number to hear a recording of the forecast was a fun novelty. (Older Gen-Xers like myself may similarly remember dialing a number that announced the time, marking the exact moment a new minute began with a beep.) By phoning 814-954-6440, anyone can hear a summary of the CTP weather forecast. Banghoff puts the call to the automated service on speaker for me. A robotic-sounding voice answers: "Welcome and thank you for calling the National Weather Service forecast office in State College, Pennsylvania . . . to hear the latest forecast for Harrisburg and vicinity, press one." Pressing 4 summons the outlook for the Lancaster area, the southeastern edge of CTP's territory and a hub of the Amish in Pennsylvania.

The automated phone service represents a lifeline for members of the Amish community. Telephones are one of the more complicated technology areas of community customs: Amish families generally will not permit a telephone inside the home but will share a community phone. These are typically located in a small shack; outsiders sometimes mistake the small structures for outhouses. The community phones are used for emergencies, for business needs—and for getting information about the weather.

Banghoff is able to fill me in on all this because one of his ongoing projects is working on weather information and safety for Amish communities. Their needs create an interesting challenge on multiple fronts. Because they minimize use of computer technology and avoid broadcast TV and radio, simple telephone calls become central to any communications plan. And because so many Amish in the area engage in farming, they have some distinctive weather interests. "It's just fascinating," Banghoff says. "This underserved, vulnerable community, are they getting the information they need? Ultimately, they've done their thing all along. They don't need us. But it probably would be helpful for us to connect with them along the way."

Given the nature of the challenge, the outreach requires a somewhat old-fashioned approach. Since emails and texts won't reach most of the Amish community members, Banghoff pulls together material for printed flyers. There's a group called the Pennsylvania Amish Safety Committee that includes representatives from the community as well as health and other officials. Banghoff plans to provide these flyers to the committee, which can then distribute them at various local safety meetings. In producing the flyers, he's able to draw on the work of other Weather Service colleagues, particularly those in the Jackson, Kentucky, office. After five Amish children died as the result of a 2020 flood in Kentucky—the horse-drawn buggy carrying them overturned on a low creek-crossing bridge, and the kids were swept away—the Weather Service launched Weather Awareness for a Rural Nation, or WARN.[72] The group produces handouts with guidance on various weather threats. Some of the flyers aren't dissimilar from what the Weather Service shares on social media, but of course, social media won't reach the vulnerable members of these communities.

Another WARN initiative seeks to reconcile the advantages of broadcast alerts with the reluctance by Amish to use radios. Across the United States, the NOAA Weather Radio network beams out forecast information nonstop. Radios with a weather-alert feature can be programmed with the user's location so that a local watch or warning triggers an alert. These radios provide a hugely underappreciated weather-safety service; well-rated models can be found for under thirty-five dollars. The traditional radio technology works regardless of whether your Wi-Fi is functioning or your cell phone has bars. Among other benefits, the emergency warning can wake you at night in case of dangers such as tornadoes. To make this tool more acceptable to the Amish, WARN partnered with a leading manufacturer, Midland Radio, to develop prototypes that take Amish concerns into account. The design strips the device down to the essential function of location-based alerts, eliminates access to regular AM and FM bands, and relies on solar or hand-crank power. The Weather Service says it hopes to see them in production soon.

Routine weather also matters for the Amish, which brings Banghoff back to the recorded forecast playing on the speakerphone. Many Amish are keenly interested in these forecasts and use them to guide decisions on their farms. Banghoff explains that the information on the automated phone service had previously duplicated what listeners heard on NOAA Weather Radio for the State College office's territory. "But the Amish would often call us wanting to know more, information about timing and amounts of precipitation," he says. One reason: hay harvesting. The old adage about "making hay while the sun shines" still applies. Hay that will be used to feed livestock must be cut and dried before storing. A three-to-four-day drying period is typical. Rain can leach nutrients from the cut hay, reducing its ability to sustain livestock over the winter, and wet hay can encourage the growth of microbes that produce heat, creating a fire risk from spontaneous combustion. The State College meteorologists realized many farmers were dialing the automated system, then pressing 9 to speak with a human forecaster with detailed questions about timing and amounts of possible rain. "We were getting literally hundreds of calls asking, 'Is it going to rain?,'" Banghoff says. "That's a huge workload." In response, the forecasters tweaked the phone recordings to include extra details. That has cut down on the number of callers needing to get through to a human, though it's still not ideal—the automated system can handle only twelve incoming calls at once, so additional callers wind up with a busy signal.

Efforts to meet the needs of the Amish underscore how the safety mission for the Weather Service increasingly goes beyond making predictions to finding ways to help ensure those predictions result in good outcomes. Underserved communities and those with special needs are especially important in these pursuits. A local meteorologist from just twenty years ago might be surprised by how the WFO is changing and how much time now goes into working with communities and organizations. For Banghoff, just a few years into his career, it's a satisfying component of his job. "Find a need, meet a need," Banghoff says. "What we do, there's such an inherent sense of purpose in it. That's the best part of the job," he adds. "Being out with people and forging relationships and developing trust. For me, that's definitely the most enjoyable aspect."

A few miles away, in room 607 of the Walker Building on Penn State's University Park campus, some potential future John Banghoffs work to hone their forecasting skills. The Walker Building, a 1970s-era cube with a Brutalist architecture aesthetic on the southern edge of campus, houses Penn State's Department of Meteorology and Atmospheric Science, widely considered one of the nation's top programs. And Walker 607, a lab-style classroom on the building's sixth floor, serves as home base for the department's forecasting practicum, better known as Meteo 415.

Classically, a practicum refers to a course that gives students a chance to take the knowledge from their studies and apply it in a practical way with supervision, such as when education majors get some experience in a classroom before becoming full-fledged student teachers. In Meteo 415, students practice developing real-world forecasts. Half the grade in Meteo 415 is based on weather-forecasting contests: Students are assigned specific geographic areas of the United States, and drawing on the technology of modern meteorology, from satellite views and radar images to computer models, they produce their own predictions and see how their prognostications stack up against the eventual real-world conditions.

This is the crux of local weather forecasting. In Meteo 415, students get more than just experience with the tools of meteorology. They also learn where the human factor comes in. They're all competing against one another to produce the most accurate forecasts, but in a way they're competing with the computer as well. When does the weather turn out exactly as the numerical models had forecast, and when do the quirks of the atmosphere, the terrain, and the models themselves throw the computer predictions off a bit? In Meteo 415, students see how experience and knowledge combined with technology produce a better forecast than the models alone.

Presiding over Meteo 415 is Professor Jon Nese, who also serves as head of the undergraduate program in the Department of Meteorology and Atmospheric Science. Nese is a Penn Stater through and through. He earned his BS, MS, and Ph.D. all right here. Earlier in his career, he worked as a meteorologist for a few local television stations, then at the Weather Channel. Then he moved into teaching. He's now approaching twenty years on the faculty at Penn State. Nese combines a gentle, patient manner with a deep understanding of both the atmosphere and the work of a forecaster. With students, he's reassuring and unhurried. I can't help but think of Fred Rogers—if, that is, Mister Rogers had a Ph.D. in meteorology.

Standing at the front of Walker 607, Nese wears a light blue fleece jacket with a Penn State logo over a blue button-down shirt, black slacks, and black Hoka running shoes. The twelve students sit at computer workstations. Two years after the start of the COVID-19 pandemic, everyone in this group is masked. Two students come up to the front of the room to deliver a PowerPoint presentation about their contest forecast. They were assigned to predict the weather around Redding, California, which is south of Mount Shasta in California's Shasta Cascade region, the northeast corner of the state.

The students kick off their PowerPoint with a few fun facts about Redding, among them that it is "home to the world's largest sundial which is also a bridge." They're describing the Sundial Bridge, a soaring Santiago Calatrava–designed span crossing the Sacramento River, and thanks to an angular support tower that casts a shadow, it really does tell time.

After the overview, the presentation gets intense. In just a few minutes, the students cycle through a tremendous amount of data they analyzed for their forecast, including a classic 500-millibar pressure chart, GOES satellite images, and radar readings. They note that the official Weather Service forecast for heavy rain "did not verify"—meteorology lingo meaning that it didn't happen. They wrap up with some lessons learned from making their own predictions for Redding, including the need to keep close tabs on the timing of any rain, since the cooling from earlier-than-expected precipitation will throw off hourly temperature forecasts.

After the students wrap up their presentations, Nese engages the entire group on takeaways from their contest experiences. "I want to start with a little retrospective," he says, putting the contest results up on the screen. He's focusing on instances in the case studies where the computer-forecast guidance from the National Blend of Models was off. The NBM combines, or blends, the output from a number of different numerical weather-prediction models.[73] The idea, as the Weather Service describes it, is that this whole is greater than the sum of its parts. NBM forecasts are a critical tool for operational meteorologists like Banghoff and his Weather Service colleagues, so interpreting them is an important skill for these Penn State students.

Walking back and forth in the front of the lab in his black Hokas, Nese draws the students out on what they've learned. Why, he asks, do they think that the actual low temperatures in their contest areas have proven warmer than the NBM predicted? A student responds that a lot of the locations for the contests have been cloudy. That's a good answer, Nese says. At night, the ground gives up heat absorbed from the sun during the day; when it's cloudy, that heat stays trapped rather than radiating back into space. "If I'm going to overwhelmingly pick places that are cloudy at night, it's going to warm up at night," Nese says. The larger point is that human forecasters need to be aware of such factors to understand when and how the NBM forecast might steer them wrong.

Between classes, I had a chance to sit down with some of the students to talk about what drove them to study meteorology and what they hope to do in their careers. Even before we started chatting, I found myself impressed. I had gotten a glimpse into the coursework needed to earn an undergraduate degree in Penn State's meteorology program: a hefty helping of math, statistics, chemistry, and physics, followed by rigorous weather-specific classes.

Here are a few samples: Atmospheric Dynamics (Meteo 421), which, among other topics, "presents the concepts of phase and group velocity with applications to gravity, inertial, and Rossby waves, and to geostrophic adjustment," according to the course catalog. There's Meteo 470 (Climate

Dynamics) and Meteo 474 (Computer Methods in Meteorological Analysis and Forecasting). Some students may opt for related courses, such as Energy, Business, and Finance 473 (Risk Management in Energy Industries), which teaches "quantitative techniques for describing how firms can use financial instruments to manage their financial risk" with a particular focus on "threats to financial viability from the weather." I was a liberal arts major, and for me, realizing the intensity of hard-core science these students need to absorb was both sobering and inspiring.

"People aren't aware of how much math and physics there is," student Alex Alvin Cheung tells me. "People think we just look at clouds all day. I have friends who have asked me, 'What's that cloud?' And I do know it, but they think that's what I study all day." Cheung is completing his BS degree, and after graduation he will head to a Ph.D. program at the University of Maryland to research machine-learning techniques and tropical cyclone predictions.

When I asked students what drew them to meteorology, the answers were striking because of how much they tracked with conversations I'd had with others in the forecasting world, from entrepreneurs to top research scientists. A significant proportion described a childhood fascination or memorable encounter with the weather. "It's a long story, so I won't bore you. But I had kind of a scary event with a severe thunderstorm," says Hunter Donahoe, a senior headed to a broadcast TV job in Texas after graduation. "High winds took off the roof of a barn that I was in. And that's pretty much how I got interested in weather." Alisha Wellington tells me about watching a TV meteorologist on the *Action News* program on ABC's Philadelphia affiliate. "I thought she was looking into the future because she was predicting the weather," Wellington says. "I thought, *I'm going to do that too.*"

The students' career goals range from private to public enterprises. Some aim for traditional careers, whether that's working at the National Weather Service as an operational forecaster like Banghoff or seeking a role as a broadcast meteorologist. Others plan to pursue Ph.D.s so they can conduct research and teach. Increasingly, some of these meteorology students see opportunities in the business world with companies that need weather intelligence to guide their decisions. Abigail McCarthy tells me she's headed into a job as a risk-management analyst for a reinsurance broker. "I didn't realize before I came here how much weather affects everything, like shipments and insurance," she says. "And I don't think a lot of the public realizes that either."

Talking with these students helped me understand some of the challenges ahead for them as well. For young meteorologists attracted to broadcasting, questions loom about local TV as digital technologies continue to reshape traditional media. Most in the industry agree that, with the growing urgency of climate change and its links to extreme weather, the need for skilled and trusted communicators is greater than ever. But the politicization of those same topics and the out-and-out climate denial among some extremists can make the friendly local weather forecaster a target—particularly on social media, where participation is a job requirement these days. (In one prominent incident in 2023, a meteorologist for a Des Moines, Iowa, TV station left his job, citing threats over his climate coverage as a factor in his decision—news that prompted other broadcasters to share experiences of being harassed.)[74] Several women in the department told me they also expect to encounter sexism, as they've watched female meteorologists get targeted on social media with comments about their looks and their clothes.

Then there are the changes underway in how forecasting works. Even as students learn in Meteo 415 the eccentricities of computer models and the value of a human forecaster's insights, they're also aware of the escalating involvement of artificial intelligence and machine learning. Several students describe the impact of their professors stressing the value of learning to code and understanding how to apply AI to weather problems. Others remarked on the need for expertise in the social sciences to make sure that good forecasts translate to good decisions and results.

No matter what their career hopes were, the students I talked with had good-humoredly resigned themselves to a future of gibes about imprecise forecasts. "It's the 'Oh, you guys are always wrong' thing," says Dana Osgood. "But if you look at the trends, the accuracy of forecasts has been increasing steadily over time."

Drive a few miles from the Penn State campus in the opposite direction from the National Weather Service WFO and you'll see a gleaming mirrored box of a building set back from the road. You'll also notice enough satellite antennas to equip a NASA mission. A few big dishes are stationed on the lawn in front of the building; an entire family of smaller dishes dots the rooftop. A cheery orange logo with a stylized shining sun announces the building's occupant: AccuWeather, a famous name in meteorology and a trailblazer of local forecasting.

The geography is no coincidence. Joel Myers, the meteorologist and entrepreneur who founded AccuWeather, is one of Penn State's most prominent alumni. Myers earned his undergraduate degree there, class of 1961, and went on to complete a master's and a Ph.D. there—all in meteorology. He taught there too. And he has maintained lifelong ties with the university; he spent decades as a Penn State trustee. That Walker Building lab where young forecasters practice their skills in Meteo 415? It's part of the Joel N. Myers Weather Center, a high-tech suite of classroom spaces, conference rooms, a student lounge area, a radio broadcast facility, and more, funded by a gift from its namesake.

Myers is, in a sense, the original disrupter of the modern weather world. While more recent weather start-ups, like hyperlocal forecaster Tomorrow.io and mini-satellite operator Spire Global, attract buzz these days for new approaches, Myers forged a distinctive path back when meteorology was far more homogeneous. He traces his own interest in weather back to age three. "I was fascinated by snow, growing up in Philadelphia," he tells me. Around age seven, he recalls, he started keeping a daily diary to track weather conditions. He set up a home weather station and embarked on a path that would take him to Penn State's meteorology program.

While working on his master's at Penn State in 1962, he had what he describes as a key insight: It was possible for a highly skilled meteorologist to out-forecast the National Weather Service—and that extra accuracy would be worth something to the right customers.

"Even my parents thought I was nuts," Myers recalls. We're sitting in his corner office at AccuWeather. Elsewhere in the building, everything looks high tech; there are walls covered in computer displays, state-of-the-art TV studios, and soundproof booths to record radio segments and podcasts. Here, though, history defines the theme. Of course there are computer and TV displays. But the walls across from Myers's desk are adorned with antique barometers from his personal collection. (He says he has more than three hundred of these forecasting relics divided between the office and his home.) There's also a poster titled "The Price of Freedom" that recounts details about the fates of the signers of the Declaration of Independence. Myers describes securing early customers that served as a starting point. Penn State professor Charles Hosler told Myers he had been contacted by a local gas company. "He said, 'Joel, I've got a guy who's interested in the high-low forecast, the five-day forecast, and he even may pay for them,'" remembers Myers, now in his eighties. It's an origin story he's told many times but one he is obviously proud of. "So I called the guy and got fifty dollars a month for three months. And we're off and running. I contacted other gas companies and finally signed up two after dozens of calls."

Sixty years later, AccuWeather is an industry fixture and a forecasting behemoth with its own cable channel. The company is privately held, so financials aren't available. In 2019, *Forbes* magazine estimated its value at as much as $900 million, a number that has likely grown in the years since. Among other services, it supplies local forecast information to TV and radio stations, which then have their own on-air talent present the forecast. If you've ever tuned in to local news in the United States, chances are you've heard a cheerful weather anchor announce, "Now, the AccuWeather forecast for our area . . ." That branding remains one of Myers's signature achievements. He says early broadcast customers were reluctant to use it, but it turned out to be a way for local stations to differentiate their weather forecasts from their competitors'.

But just as Myers pioneered the sale of meteorological services, he has also stoked some controversial debates about the balance between private and public forecasting. To understand these tensions, you need to know more about a term that I heard over and over again in all parts of the US meteorology community: the *weather enterprise*. As I've mentioned elsewhere, it's shorthand for a concept that describes the relationship—and, ideally, the synergy—among three primary groups in the weather world. The first is the US government, which funds NOAA and the National Weather Service, launches big satellites, operates the nationwide network of radar sites, and runs critical weather models on its supercomputers. The second is the private sector—AccuWeather and others—which sells forecasts and other services, generally filling niches too specialized for the government. The third is the academic world, including campuses like Penn State, where top scientists conduct research and train the next generation of weather experts.

The interplay of the legs on this tripod is generally regarded as beneficial—indeed, as a foundation of US success in meteorology. In March 2023, Antonio Busalacchi testified before the Environment Subcommittee of the House Committee on Science, Space, and Technology. Busalacchi is the president of UCAR, the University Corporation for Atmospheric Research, a nonprofit that operates the National Center for Atmospheric Research, a federally funded research center. In his testimony, he reviewed the concept of the weather enterprise and implored lawmakers to ensure that its three big components remained in a productive balance. "It is important to the future success of the weather enterprise that each leg of the triad continues to grow, and that any contraction or reduction in size of any leg will negatively impact its diverse beneficiaries," he said. "As extreme weather events become more prevalent and costly, it is critical that innovation for prediction keep pace with our changing world."[75]

But the general happy-talk tone of the weather enterprise discourse masks real tension about where government, academia, and the private sector compete with rather than complement each other. From time to time, that tension flares up. One of those eruptions took place in 2005, with AccuWeather in the center of it. Rick Santorum, who at the time was a Republican senator from Pennsylvania, proposed legislation mandating that the National Weather Service limit its forecasting primarily to "preparation and issuance of severe weather forecasts and warnings designed for the protection of life and property of the general public" and requiring the government to refrain from providing a "product or service [. . .] that is or could be provided by the private sector."[76]

Santorum's proposal was immediately blasted from multiple directions. It would have effectively taken Banghoff's predecessors in the State College WFO and their colleagues across the United States out of the work of sharing local forecasts, limiting them to issuing severe-weather warnings. Critics pointed out that a prime beneficiary of this setup would be AccuWeather—and it didn't help that Joel Myers and his brother Barry, at the time also an AccuWeather executive, had contributed to Santorum's campaign and the Republican Party, as the Associated Press and others reported.[77]

Santorum's proposal, critics said, would have taxpayers footing the bill for collecting the meteorological data while letting companies like AccuWeather reap the benefits by selling forecasts. At the very least, it raised public-safety concerns about everyone having equal access to information about hazardous weather. The Santorum bill went nowhere, but it dramatized the concerns about privatization.

AccuWeather generated controversy again in 2017 when Donald Trump nominated Barry Myers to lead NOAA, the parent organization of the National Weather Service. In the context of NOAA's history, Myers was an unusual choice. Unlike his brother Joel, Barry was not a scientist; he had degrees in business and law. Advocates argued that Barry Myers could bring some much-needed entrepreneurism to NOAA. But AccuWeather's previous advocacy for more privatization raised fears about conflicts of interest. Barry Myers's nomination wound up damaged by revelations that AccuWeather had paid $290,000 as part of a settlement "after a federal oversight agency found the company subjected female employees to sexual harassment and a hostile work environment," according to a *Washington Post* report in February 2019.[78] AccuWeather denied the allegations but agreed to a number of responses in the settlement. Barry Myers eventually withdrew himself from consideration for the NOAA post, citing health concerns.

Some of this public-private tension can't be avoided. Even many big supporters of NOAA and the National Weather Service agree that these agencies need to be more nimble, particularly given technological changes underway, from new types of satellites to the growing role of artificial intelligence. Private companies in those areas are demonstrating real advances, and it's important to encourage innovation. At the same time, weather affects us all. The public-safety mission of the National Weather Service benefits everyone—including companies like AccuWeather, which feed taxpayer-funded and government-produced data into their own forecasting systems.[79]

When Joel Myers recounts the history of AccuWeather for me, he's unapologetic about pushing for free enterprise even as he acknowledges the need for a government role—in his view, mainly as a data source and to issue public-safety warnings. "Their job is to save lives and issue those warnings," Myers says. "And most importantly, to gather the data. Because the data is the basis of what goes into the computer models, and that serves all entities, public, private, and academic."

Downstairs from Myers's office, AccuWeather meteorologists operate from a cavernous space dubbed the Global Forecast Center. Jon Porter, AccuWeather's chief forecaster (and a Penn State alumnus, unsurprisingly), shows me this operations center from the mezzanine level before taking me down onto the floor. The entire room has an open-plan setup—there are rows of workstations and a high ceiling that allows for setting up lights and cameras at individual desks when needed; everyone can see the large banks of monitors on one wall. Porter says the design was intended to foster collaboration. "We know we're producing the best forecast when we're working together," he says.

As we make our way across the ops center floor, staffers drift toward a cluster of workstations for a meeting. Though I'm not a meteorologist, as the discussion unfolds, I can relate as a journalist: it's essentially a news meeting, like so many I've participated in at newspapers and magazines. The group quickly talks through areas of the United States likely to experience severe weather. Together they review a graphic headlined "SEVERE T-STORMS / Wed-Wed Evening." It's a map showing a big red blotch centered over Oklahoma City and extending south to the very top edge of Texas and north to just touch Kansas City. Bullet points list possible dangers: localized flash flooding, large hail, isolated tornadoes, and damaging wind gusts. After reviewing internal forecasts as well as updates from the National Weather Service, one person asks: "Anyone see a reason to change the avocado?" (The red blotch, I realize, is in fact shaped like an avocado.) A colleague who has dialed in through a conference call chimes in that the current hazard area might extend too far north: "We might not need to have it on Kansas City," he says.

That conference call connects the Global Forecast Center here in State College with another AccuWeather facility in Wichita, Kansas. Meteorologists there specialize in custom forecasts for business clients who want to be alerted about weather likely to affect their facilities or operations. One example that AccuWeather likes to highlight concerns a February

2008 tornado in Oxford, Mississippi. At the time, the city was home to a factory of Caterpillar, the construction-equipment giant. According to an AccuWeather history of the incident, its forecasters issued a warning to the plant at 5:37 p.m., when the storm was eighteen miles away—notably, AccuWeather says, before any other warning from any other weather agency. Plant officials moved workers into an emergency shelter. When the tornado hit at 5:59 p.m., it ripped through the factory, but no one was injured. The workers were all safe in the shelter.

"We forecast for our clients and what the public needs," Myers says. "And so we believe that our forecasts have much greater accuracy when you talk about their impact and the decisions." AccuWeather says it draws on a broad base of forecast data and multiple computer models for its predictions, going beyond Weather Service and NOAA information to include private sources, then using its own proprietary forecast technology. Myers says he's proud that AccuWeather often tops accuracy ratings for forecast companies but adds that accuracy is more complicated than just hitting a number. "Let's say a competitor predicts a high of thirty-five, mostly cloudy, for tomorrow. And AccuWeather predicts thirty-four in the morning, with temperatures falling sharply through the twenties into the teens in the afternoon, with all water and slush freezing up." If it turns out the high was 35 degrees and not 34, Myers says, technically the competitor won. But, he asks, which forecast was more useful? "We told the story."

At 9:02 p.m. at the State College WFO, John Banghoff begins to focus on updating the long-range forecast for central Pennsylvania, a component of the area forecast discussion he will release before wrapping up his shift for the day. Over the course of hours, I've watched as Banghoff has methodically worked his way through a detailed checklist for the meteorologist handling the public-service duties on today's 3:30 p.m. to 11:30 p.m. shift. He marks each item as done in a shift-log spreadsheet. The two other forecasters on duty—one on the short-term weather shift, the other on the medium- to long-range shift—are hunched over their workstations across the bullpen. After a brief negotiation, it's agreed that Banghoff will take on the update of the long-term forecast while his colleague catches up on other tasks. It also gives me an opportunity to watch the core responsibility of an operational meteorologist: reviewing all the available information and producing a forecast.

At this point, Banghoff has been on shift more than five hours. After creating the Weather Story, he pulled together a separate version of the outlook, referred to as the briefing. The Weather Story summarizes some key information for the public, but the briefing goes to a different audience. "This is designed specifically for emergency managers and what we call core partners," Banghoff explains. "It's got more detail. It goes into more depth on what we're expecting." The WFO sends the briefing directly to those emergency managers and public-safety officials as well as to other administrators who may find the alerts helpful: school superintendents, managers at the state's Department of Transportation, hospital officials, key staff at infrastructure organizations such as utilities and public works, and more. At this moment, the distribution list numbers 403 people. As it was with the Weather Story, the briefing theme today is heat. The insiders who get the briefing already know basic preparedness measures for high heat, so Banghoff's goal is alerting them to the expected high heat index values, which could lead to a heat advisory. "Just so we keep messaging that possibility," he says while typing information into a template.

Here's a sampling of other duties that Banghoff has been dealing with over the past several hours. He consults a list of spot-forecast requests from agencies and outside groups that have asked the WFO to monitor the weather ahead of an event or for a specific location. These requests can range from local groups planning county fairs and other outdoor events to forestry officials plotting a controlled burn to prevent a future uncontrolled wildfire. Satisfied that there are no outstanding requests, Banghoff moves on to verifying that the automated phone system is updated and functioning. He checks the State College WFO social media accounts to see if any Facebook or Twitter users have left comments in need of a reply. He works on the flyers for Amish weather awareness and places an order to get them printed up. All the while, he keeps an eye on the splash of windows open on his AWIPS display to track the actual weather. Along the way, we both eat a quick dinner at his desk—Banghoff nibbling from a meal he packed at home, me enjoying a burrito bowl brought to the WFO by DoorDash.

At one point, Banghoff turns to another key responsibility of every Weather Service WFO: reviewing climate data. These are the weather observations collected by the network of Automated Surface Observing Systems, or ASOS. The ASOS weather stations are sited mainly at airports across the United States, providing critical information on conditions for aircraft taking off and landing but also taking measurements that become the official records of the weather. If you visit the website for your local WFO, you can see these climate reports. A daily summary includes the location's high and low temperatures and the times those were recorded, wind-speed averages and highs, humidity highs and lows, and more.

These kinds of observations trace their history in the United States back to Thomas Jefferson. His archived papers show that he kept a daily record of temperature and conditions, beginning at least as early as July 1, 1776, while he was in Philadelphia for the Second Continental Congress.[80] Basic climate data can provide fun factoids. For instance, the lowest temperature ever recorded in New York's Central Park was negative 15 degrees on February 9, 1934 (based on observations going back as far as 1869).[81] But they also represent a key scientific resource and historic record, all the more valuable as scientists try to understand and monitor climate change.

What I find interesting when Banghoff reaches the climate-data task on his checklist is a duty he refers to with two letters: QC. Depending on your own work, that shorthand may be familiar to you. It stands for "quality control." The responsibility of QC came up throughout Banghoff's time on shift. With so much information collected and generated automatically by computer systems, it's the job of humans in the loop to look for anomalies that could indicate a momentary glitch or equipment in need of calibration.

The State College WFO monitors climate information at five central Pennsylvania stations: Altoona, Bradford, Harrisburg, Johnstown, and Williamsport. Some have records dating back to the nineteenth century. "I basically just QC to make sure everything looks okay," Banghoff says. He notes one of the stations recorded a trace of precipitation, but he knows that there wasn't actually any rain. The reading probably reflects earlier mist. "So that's going to be a false trace of precipitation because no precipitation actually fell," he says, correcting the data. Thanks to the National Weather Service, we have complete and consistent records. And no matter how automated things get, a human in the loop—even doing just a quick check—ensures we can treat that data as reliable.

After ticking off more checklist items, Banghoff turns to the long-range forecast, which he needs to complete the area forecast discussion before it can be sent out. In the context of a WFO's local forecast, *near term* is today's forecast, *short term* covers days two and three, and *long range* describes days four through eight or more. Since this is a Tuesday night, for the long-range forecast Banghoff is looking at Friday through the weekend and into early next week. Once again, Banghoff flicks through different data sources and displays faster than I can process them. I count twelve different windows open at once on his AWIPS display and other screens, including barometric-pressure maps produced by the computer models and temperature predictions. Banghoff knows the outlook from the previous long-range forecast issued by the WFO, so he's watching for indications that the latest computer models have predicted something new or different.

Once more, there's some QC involved. The computer system can populate the forecast template with temperature and precipitation predictions directly from the computer models. "So the first one I'm going to run is that my sky [cloud] coverage always has to be greater than or equal to my probability of precipitation," he explains. "In other words, I can't have a sixty percent chance of rain if I don't have at least sixty percent cloud cover, right?" Another example: The hourly temperature forecast numbers must always fall between the predicted maximum and minimum temperatures.

Finally it's time to summarize all this information for a synopsis of the long-term forecast. As Banghoff works, I'm reminded of my early reporting experience writing stock-market wrap-ups for Dow Jones and *The Wall Street Journal*. Just as with those end-of-day market summaries, WFO products have their own particular cadence and vocabulary. Banghoff speaks the words aloud as he composes and types: "Southwesterly winds will pump in increasing moisture and warmer temperatures, with heat index values again approaching the century mark across much of Southeast Pennsylvania by Sunday." He keeps typing, noting "increasing chances of showers and thunderstorms from northwest to southeast as we progress through the weekend and into early next week." Part of the forecaster's job is to communicate the possibilities and uncertainties. "Dry conditions should prevail through most of Saturday for all but the northwest corner of the Commonwealth, but showers and storms should be more common by Sunday afternoon and evening," Banghoff writes. This later proves to be the case. The weekend turns out hot and humid and includes scattered severe thunderstorm warnings in parts of the WFO's coverage area on Sunday.

As Banghoff works through all the forecast data, we talk about how the computer models keep improving. It's going to keep getting harder for those aspiring mets in Meteo 415 to out-forecast the computer. AI and machine-learning advances will make the automated forecasts even more reliable. Does Banghoff, I wonder, see himself being effectively replaced by a computer in the years ahead?

"As the data feeds and speed exponentially increase, humans will quickly lose the ability to be effective at parsing through all that data and making good decisions," he says. "Now, thirty or forty years ago, before my time, when there was a radar screen coming in every fifteen minutes, that's pretty easy for a human to process and make decisions on. But now we have radar data from potentially three or four radars coming in every minute, combined with satellite data coming in every thirty seconds, with multiple different products. There's an exorbitant amount of data that's coming in.

"But I think the role of the human is going to continue to be super-relevant," he adds. "I think it's just shifting a little bit. In the past, the local TV meteorologist was the primary source for weather information, right?" Those broadcasters, often working from Weather Service predictions, could easily convey whether the outlook was routine or dangerous, something that might not come through as clearly on a smartphone app.

"So I think what's going to be critically important moving forward is that a meteorologist is not only a communicator but a salesman to some extent," Banghoff says. He says he hopes the automation will, rather than replace him, make routine forecasting more efficient, opening more room for those other duties. "The role of the meteorologist, I think, will be to be creative about getting the word out to people. And, when it's absolutely necessary, to sort of scream and say, 'Hey! You really need to pay attention.'"

4 Hyperlocal Weather

THE NEW POSSIBILITIES OF ZOOMING IN TIGHT

On a sunny spring afternoon in Virginia Beach, I'm waiting on a package delivery from Walmart. That probably doesn't sound too remarkable. Ever since the rise of e-commerce, and particularly after the COVID-19 pandemic began, ordering online and getting goods dropped off at your home has evolved beyond convenience into a trillion-dollar segment of the US consumer economy. Indeed, according to the Census Bureau, in 2023, e-commerce sales totaled 15.4 percent of all US retail sales, more than two and a half times the share a decade earlier.

But today, I'm not looking for a truck or the neighborhood postal carrier. I'm watching the sky. Here in Virginia Beach, with Naval Air Station Oceana less than a dozen miles away, it's not unheard of for an F/A-18E Super Hornet fighter to streak overhead. I am squinting to locate something much smaller. I begin to hear a muted buzzing that ramps up in volume. Then I see a small dark dot in the distance. A few seconds later, it resolves into a shape: a white boxy case with two orange bulges on top, two dark struts projecting down, and four projecting out from the center. The tips of the four projecting struts are blurry circles. As it nears and the buzzing gets louder, the object reveals itself clearly: a quadcopter-style drone moving steadily overhead. Seen from a distance, it's not unlike the small consumer drones you may have spotted people flying in their backyards or at the beach.

Then this drone does something I've never seen before. While the craft hovers above nearby trees, a smaller box detaches from underneath it. At first the box looks like it's falling, but then I'm able to make out a thin line connecting it to the drone. The box continues to descend and then, with surprising gentleness, touches down on the grass near where I'm standing. A bulbous red grappling device separates from the box and shoots back up, pulled by the retracting tether. Throughout this process, the drone remains placidly in place directly over the drop site. The charcoal-gray box left behind sports an orange-and-white logo for DroneUp, the start-up that's sending these packages flying through the air.

The shipment I've just received came from Walmart, but it feels like a delivery from the future. The box brought by the drone is a cardboard container with a handle formed by the folding flaps on top, a larger version of the type of carry box you'd see at Dunkin' if you bought a couple dozen Munchkins. This particular box contains bottled water, a test payload favored by the company behind this new way for us to get our goods, but the drone could have brought me cookies, ice cream, shampoo, aspirin, or any of hundreds of qualifying goods from the nearby Walmart Supercenter.

You may have heard or read about drone delivery. In some parts of the United States—including the Virginia Beach neighborhood where I'm standing—it's already here. Advocates say that using drones, often referred to as UAVs (traditionally the acronym stood for "unmanned aerial vehicle," though lately it's being updated to the gender-neutral "uncrewed aerial vehicle"), can help reduce emissions, unclog roads, and provide speedy service to get goods into customers' hands. DroneUp and other companies constitute a burgeoning industry sprinting to make this happen. These efforts, though still early, have already advanced significantly in just a few years. But this future depends in no small part on weather intelligence to assist in planning and to ensure safe and efficient flights.

"The saying goes, there's better, cheaper, and faster, and you get to pick two of those," says John Vernon, cofounder and chief technology officer at DroneUp. "But we're entering a space here with UAVs where better, faster, and cheaper are all realistically feasible and within our grasp. I know that sounds a little bit crazy. But this is probably one of the least expensive ways, ultimately, to deliver goods."

Vernon walks me through each of those aspects: *Better*, he argues, because drones are typically battery-powered. If operators use electricity from renewable sources, their drones will contribute to carbon reductions. *Faster*, Vernon says, because drones don't need to restrict themselves to convoluted roads; they can take direct routes—go as the proverbial crow flies. *Cheaper*, Vernon tells me, because of continuing advances in automation, UAV designs, battery technologies, and other factors. And they're safer, he adds. "If you think about road accidents and that kind of thing with traditional transportation, we're creating an environment where a lot can be taken off the roads."

DroneUp is already proving that flying deliveries can work for Walmart, the biggest retailer in the United States. The start-up, like others pioneering everyday applications for UAVs, has bold plans for the future. But every flight takes off at the whims of the weather, particularly the winds. Gusting air can affect an uncrewed craft's stability. Strong headwinds force a drone to perform the aerial equivalent of swimming upstream, depleting its batteries at a faster clip. "Weather has a huge impact," Vernon tells me. "I would say twenty-five percent of our current downtime is due to weather-related issues."

To bring about a delivery revolution, Vernon and his competitors need good weather forecasts. In fact, they—along with pioneers in other emerging industries—need a special kind of forecast. They need hyperlocal weather.

Local forecasts tell you if your city will see rain or snow or high heat, but hyperlocal weather aims to provide current and expected conditions at a far more granular scale. It's a daunting challenge and one that has only recently become practical due to advances in computing power, local observations, and machine learning. At the same time, demand for hyperlocal is growing. It isn't only start-ups like drone-delivery companies and self-driving-vehicle makers who hope forecasts can enable their new industries. More traditional companies are discovering that they can realize efficiencies and make more money when they have a better handle on what nature is about to throw at them. Consumers can access some of this kind of forecasting thanks to smartphone apps like Apple's Weather, which promises "down-to-the-minute" predictions for when it will rain at your location and comes free on iPhones, and similar offerings from the Weather Channel, AccuWeather, and others. But the real action centers on developing forecasts that can be sold to paying customers, often with weather information integrated with other tools and data targeted to specific industries.

How hyper is hyperlocal? To get a sense of that scale, consider some of the computer weather models out there. NOAA's Global Forecast System runs at a resolution of thirteen kilometers, or about eight miles, to generate a worldwide outlook.[82] The model used by NOAA to zoom in more tightly on the United States, the High-Resolution Rapid Refresh model (HRRR, usually pronounced "hurr"), operates with a three-kilometer, or 1.8-mile, resolution.[83] Contrast those with the Comprehensive Bespoke Atmospheric Model, or CBAM, developed by Tomorrow.io, a hyperlocal weather start-up. Tomorrow.io says the CBAM can be run at resolutions as small as tens of meters—a scale that can effectively predict how the weather will differ from one city block to another.[84]

Of course there are trade-offs. NOAA's GFS models the weather across the entire Earth; the coarser resolution makes it feasible to complete the calculations needed to cover all that territory in a reasonable time. Meteorologists depend on the GFS to look further out, up to sixteen days in advance. Higher-resolution models sacrifice that scope in distance and time to zoom in to hyperlocal scales.

Hyperlocal observations can help drive improvements in hyperlocal forecasts. Researchers and companies working in the space hope to tap a number of sources beyond traditional satellite and radar. One is the Internet of Things—the billions of internet-connected devices that gather data, from barometric pressure sensors built into common cell phones to webcam video feeds whose images can be analyzed to determine current weather conditions. Across the country, states have been building networks of weather stations called mesonets that can supplement the National Weather Service monitors typically found at airports. And LIDAR technology—which can measure wind speeds at a variety of heights by bouncing lasers off particles in the air—can gather information from a small box on the ground, generating observations that would previously have required a tower with an anemometer.

Nearly every segment of weather forecasting has some mix of public and private organizations. Indeed, as you've seen elsewhere in this book, the concept of the weather enterprise—that triad of government agencies, research universities, and high-tech companies that interact to advance forecasting—is nearly ubiquitous in any US policy or capability discussion around meteorology. But hyperlocal is one area where the efforts of private companies are growing more prominent. That's at least in part because business customers have the most to gain from hyperlocal predictions, which means go-getting companies see a market in providing them. It isn't that the National Weather Service doesn't provide services valued by businesses. NOAA and the Weather Service are part of the Department of Commerce, and the mission statement for the Weather Service invokes "the protection of life and property and enhancement of the national economy."[85] But the Weather Service isn't in a position to develop niche solutions for hundreds of different industries at the taxpayers' expense.

And so private forecasting companies large and small are driving these advances. Microsoft, for instance, deploys its machine-learning technologies to refine weather predictions, which it then incorporates into broader cloud-computing products it markets to growers and agriculture companies integrated with other features to track data about everything on the farm. Tomorrow.io provides hyperlocal forecasts to major airlines to help guide airport operations and administrative decisions. Meteomatics, based in Switzerland, develops drones that collect atmospheric measurements while they ascend and descend, feeding local data into weather models. Tempest sells easy-to-use home weather stations that effectively crowdsource hyperlocal observations.

If you're thinking it would be great to know what time rain will arrive at your kid's soccer field for the championship game two weeks from Tuesday, don't get too excited. Despite the obvious attraction of knowing the weather for your neighborhood block by block, truly hyperlocal forecasts remain primarily a specialty service best employed by users with unique needs (individual business owners, small companies, and big corporations), a sophisticated understanding of the uncertainties involved, and a willingness to pay for better intelligence.

But the potential will only grow as forecasters amass more local data and machine-learning systems identify the atmospheric quirks of various locations and as organizations willing to pay for this kind of weather intelligence figure out the best ways to put it to work. Whether you're flying drones, designing cars to drive themselves safely during all conditions, choosing which field to fertilize tomorrow, or making any of hundreds of business decisions that can be affected by the weather, hyperlocal promises a new way to think about forecasting.

"I care about the temperature at the ground," Andrew Nelson tells me. "Not thirty feet up." Nelson runs a farm in Washington State. He's explaining to me all the ways that weather affects his crops and all the ways that weather forecasts inform the decisions he makes that ultimately affect the farm's bottom line.

Temperature, for instance, helps determine the most effective time to spray for weeds. (Apply when it's chilly, and the offending plants will absorb the herbicide too slowly for it to be effective; hotter conditions can cause the weed killer to evaporate before it can soak in.) But unnoticed by most people—because it doesn't make enough of a difference to get noticed by most people—the temperature forecasts created by National Weather Service computer models reflect predictions for ambient temperature (air temperature) two meters above the ground.[86] That's about six and a half feet. "Surface" wind forecasts are actually made for ten meters above the ground, or about thirty-three feet.[87] All this works fine for how we humans experience the weather. After all, our heads are usually five or six feet up, not down at ground level. But those weeds Nelson worries about live in the dirt. As farmers know, frosts occur even when the forecast temperature is above freezing because the soil can be several degrees cooler thanks to sinking cold air and the rate at which the ground gives up heat accumulated during the day. Variations in terrain can further complicate the picture.

The Palouse, the region of the northwestern United States where Nelson grows his crops, can itself be complicated—at least meteorologically speaking. It's also beautiful. On gently rolling hills, sunlight reflects off lush greens in the spring and sets yellows and golds aglow in the autumn. The Palouse spans parts of southeastern Washington as well as portions of Idaho and, some say, Oregon. This is some of the most fertile land in America due to mineral-rich loess soils that were created by glacial movement and blown by winds to form the hills.[88] Fields here produce soft white wheat, a lower-protein variety that's ideal for baking pastries and cookies. But the rises and falls of the hills mean weather conditions can vary between two points located relatively near each other.

"That terrain makes it very difficult, with the microclimates in our area," Nelson says. "The forecast may call for forty degrees. Problem is, in the lower areas, in the draws and drips of our fields, that can mean below freezing." This is exactly the kind of difficulty that hyperlocal weather forecasts can address.

Nelson grew up on this land. He's a fifth-generation farmer; he's raising his family here and working with his parents. He combines that agricultural DNA with the soul of a techie. Nelson earned a degree in computer science and a business degree at the University of Washington. After graduating, he worked for a few years in Seattle at consulting companies. But ultimately, he returned to the Palouse. "I wanted to apply technology to our farm," he says. He recalls an old system of paper tickets to track truckloads of grain reaped from the fields. "I think it was my freshman year in college. I was walking back to the house with that day's tickets, a little whirlwind went by, and we literally lost like five or six of the tickets," he says. He moved the information to a spreadsheet and eventually set up smartphone apps that farmworkers can use to track the harvest.

"The thing is, I keep tweaking to get more and more accurate, which is kind of how I approach everything," Nelson says. That includes forecasting the weather. Nelson started out by operating his own weather station on the farm. Then he met Ranveer Chandra. And then things started getting very hyperlocal.

Ranveer Chandra is a veteran researcher and executive at Microsoft. He's a Cornell computer-science Ph.D. with a history of internet innovations and a passion for putting technology to work in new ways. In 2015, he launched a Microsoft program called FarmBeats to explore how technologies like artificial intelligence and the Internet of Things can transform agriculture.[89] He talked with me about what he considers the stakes for smarter farming.

"Agriculture plays such an important role in climate change," Chandra says. "It's a significant emitter of greenhouse gases."[90] Farmers, Chandra says, will be among the most affected by climate change. "Especially in emerging markets, where the farmers are not as technology-equipped," he says. "Any change in weather is going to impact them the most because they just can't adapt to it. But agriculture can also be a solution to climate change. If done right, it can put more carbon back in soil."

As Chandra and his colleagues considered these problems, meteorology came into focus as one strategy. "We realized that weather plays such an important role in everything," he says. "In farmers' decisions, it helps in climate adaptation, it helps even with carbon sequestration. So that is what led to this work on microclimate prediction. The weather on the farm could be so different from the weather station that's miles away. In emerging markets, it's worse when we don't even have historical weather station data."[91]

Andrew Nelson and his family farm became a testbed for Microsoft's technologies. The work proved more complex than just firing up some computers to crunch weather-model numbers. Chandra wanted to tap the power of artificial intelligence and the ability of machine-learning systems to analyze vast amounts of data and train algorithms to make predictions. But machine learning depends on having data to teach the model.

Ideally, Nelson would have years and years of observations from points all across his land. Unfortunately, creating a hyperlocal network of weather stations has some challenges, particularly in farmland. Lack of broadband internet access remains a notorious problem for rural communities. (Government initiatives at multiple levels, including a Department of Agriculture program called ReConnect, have sought to address this by providing grants and loans to build out broadband in these areas.) "You're limited on wireless range. We have connectivity issues," Nelson says. He considered weather stations that would use cell-phone connections to relay data, but those were expensive and suffered from poor cell reception in rural Palouse. Wi-Fi networks have limited range, requiring placement of routers and repeaters. Working with Microsoft and the tools they were testing with FarmBeats, Nelson found a better approach: a wireless broadband technology known as TV white space, which taps the unused frequencies (the white space) in between television broadcast channels.[92] "We placed the weather stations around the farm. We started out with five or six," Nelson says. "Now we have over twenty of these TV white-space weather stations across our farm."

By traditional AI methods, Chandra says, it would have taken quite some time to amass enough data from Nelson's sensors to train a prediction model. But the Microsoft team realized they could take a shortcut using the robust archive of data from established weather stations, such as those operated by NOAA and the National Weather Service. The approach combines those traditional sources and models with the readings taken at the farm to create the customized hyperlocal forecast. Microsoft called this model DeepMC, a nod to the so-called deep-learning AI technique involved (the MC stands for "microclimate"). As an academic paper from Chandra and his colleagues explained, rather than predicting the weather directly, the DeepMC approach predicts the expected difference between the traditional forecast and conditions on the farm.

The results—comparing the DeepMC forecast against the predictions from Dark Sky (a hyperlocal forecast service that has since been bought by Apple and integrated into the iPhone Weather app) and the actual conditions recorded by Nelson's sensors—were impressive. In one example cited in the research paper, DeepMC correctly predicted below-freezing temperatures when the traditional forecast did not. This guided Nelson to hold off on applying fertilizer, which loses effectiveness when the ground is too cold. "Had the farmer relied on weather station forecasts [. . .] he would have been at risk of endangering the crop losing up to 20% in yield," the Microsoft team wrote.[93]

Microsoft continues to evolve its AI-powered tools for farmers through its Azure cloud-computing platform, and Nelson looks forward to future advancements. He believes hyperlocal predictions will get even better with the development of low-cost, battery-free temperature and soil sensors that could be deployed much more extensively on farms, gathering ever more data. And though he now has high-speed fiber internet at his farm, Nelson hopes other rural areas will see more broadband access, a requirement for fully reaping the benefits of AI and cloud technologies.

Chandra says he expects the value of microclimate forecasting to keep growing, particularly as climate change shakes up what's normal for any given parcel of land. "Weather and climate impact everything around us," he tells me. "Everyone is conscious of it in every industry I work on, from retail to financial services and insurance. Everyone is going to be affected by these weather predictions, and we need them to be accurate."

In the Okere District of Ghana's Eastern Region, halfway between Lake Volta and the capital city of Accra on the Atlantic coast, Mary Okainbea knows well how weather predictions can affect everything. Okainbea, a mother of five in her early fifties, farms a small family plot. She grows mostly maize—better known as corn in North America—along with some cassava. Money earned from selling the maize, along with income from her husband's job laboring on another farm, supports the family.

"I was born into farming," she tells me with the help of a translator over a video call set up inside a church in the town of Nkurakan. She adds that she dropped out of school when she was young, so the farming skills she learned from her parents are her only trade. With her husband employed elsewhere and her own children in school, Okainbea explains, she does nearly everything herself. She plants. She sprays. She harvests. "It is a lot of hard work," she says.

Recently Okainbea discovered a way to greatly improve the yield from her maize crop thanks to a hyperlocal weather forecast. This weather intelligence comes from Ignitia, a Sweden-based start-up. Predicting rainfall in the tropics can be especially tricky, and unexpected rain at an inopportune time can strike directly at a farmer's bottom line.

Okainbea walks me through an example. Her farmland lies on a hillside. If she applies fertilizer to her crops just before a storm, the rains will wash it down the hill and away from her maize. When that happens, it's a double disaster. The money spent on the fertilizer winds up wasted, and improperly fertilized crops won't be nearly as bountiful as those that received the right nutrients at the appropriate time.

That dilemma receded once Okainbea signed up for a pilot program in her region and gained access to Ignitia's forecasts. Each day, she received an SMS text message on her cell phone from Ignitia. She held up her cell phone on our video call so that I could see. It was a flat black rectangle with a small screen and a physical keypad, a design I have rarely seen in the United States since the early 2000s.

The Ignitia team knows that many of the farmers who are potential users don't have smartphones and rely on app-free, non-web-browsing models like the one Okainbea has. So the company keeps the information simple, with text messages like "Today, rain likely. Tomorrow, likely dry." It also beams out longer texts with advice on good agronomic practices. For example: "Do the first weeding from 1–2 weeks after planting to remove weeds and loosen the soil to support plant growth."

Okainbea explains that because she never learned to read, she depends on family and neighbors to report the texts to her. "As soon as the sound of the message hits the phone, if there's nobody in the house, I run out to look for whoever can read for me and guide me on what to do," she says. If the forecast calls for rain, for instance, she won't spray pesticides and will focus on another activity, such as weeding by hand.

What's the payoff? Okainbea says she performed her own experiment when she first signed up for the Ignitia texts. On one acre of her farmland, she followed the weather forecasts and agriculture advice from Ignitia. "I wasn't too sure of what the information would be like," she says. On another acre, she relied on her typical practices and instincts. On that acre, she says, she eventually harvested three standard sacks holding one hundred kilograms of maize each, a total yield of three hundred kilograms. But on the other acre—the one where she followed the Ignitia guidance—her yield was five sacks, or five hundred kilograms.

In other words, Okainbea *nearly doubled* her yield thanks to weather intelligence and agriculture-science advisories. That's not as unbelievable as it might seem. In the United States, where farms deploy every conceivable practice and technology, the average yield of corn per acre is something like nine times Okainbea's big success. But for Okainbea, that yield increase translates to a substantial monetary gain. Though commodity prices in Ghana fluctuate just like anywhere else, that year a hundred-kilogram bag fetched five hundred to six hundred Ghanian cedis, according to farmers in the region. At the exchange rate at the time, that translated to roughly $80 to $100 per bag—meaning the yield increase of two sacks on her experimental acre represented up to an extra $200 of revenue before costs, no small gain in an economy where the per-capita gross domestic product is under $2,500. "It has increased my income," Okainbea says proudly.

Ignitia, the company behind the forecasts that Okainbea receives on her phone, was launched with a goal of improving weather forecasts for the tropics and bringing those improvements to farmers throughout the region. Andreas Vallgren, Ignitia's chief science officer, says he had always been interested in meteorology. But it was after his undergraduate studies, when he participated in the United Nations peacekeeping mission in the Democratic Republic of the Congo as a meteorological observer, that he started to focus on the problem of tropical meteorology. Eventually, after completing a master's degree in meteorology and a Ph.D. in fluid mechanics, Vallgren cofounded Ignitia.

"We had an idea in mind, and that was to bring science to underserved people," Vallgren says. "West Africa has more than fifty percent of the population engaged in farming, and it's mostly smallholder farms. Few have access to irrigation, so they're dependent on what falls from the sky."

Tropical forecasts have distinct challenges compared with those for the mid-latitudes (which include most of North America and Europe). Several factors contribute to this difference, including the greater amounts of direct sunlight over a year near the equator. Rainfall predictions can be especially tough. In the mid-latitudes, a significant amount of rain comes from large-scale weather systems. But in the tropics, convection—rising masses of warm, moist air that cool as they get higher—drives more of the rain. These convective storms are inherently smaller scale and more difficult to forecast. Vallgren and his colleagues developed a regional computer model with high resolution to capture convection.[94]

But as with other types of weather around the world, accurate predictions are one thing; doing something about the weather—getting that information to the right people and in a helpful format—is another. "Most farmers in West Africa, they don't have access to smartphones, they use cheaper phones," Vallgren explains. "That's the big reason we are still using SMS." The SMS, or Short Message Service (the standard for basic text messaging), system also allows Ignitia to receive subscription fees in the form of micropayments billed through users' wireless accounts each time a text is received at a cost of a few US cents a day. By 2022, Ignitia says, it already had more than two million smallholder farmers signed up for the service. The text updates are kept short, simple, and actionable. Users with smartphones or web access can get more detailed versions of the forecasts. That appeals to owners of larger farms as well as government extension agents who work with and advise local farmers.

Ernest Amanor-Larbi is one of those extension agents. As a government employee, he helps local farmers be more successful. (He also helped set up our video call along with Ignitia's local agent, Miranda Osei Agyemang, so that I could talk with some of the farmers.) He tells me he's responsible for nineteen communities in the region. He worked with Ignitia in rolling out the text messages about the weather and the best practices for sowing and spraying. He hopes to see these services continue to expand. "It helps the farmers and also helps me do my job as well," he says. "It's reaching all of these farmers every day."

Of the many weather start-ups working in hyperlocal forecasts, one of the most audacious is Tomorrow.io. Its leaders believe the market for weather intelligence can only grow. "Weather forecasting becomes more important because of climate change," says Shimon Elkabetz, the company's CEO and cofounder. "This is already the era of consequences."

Tomorrow.io—pronounced "tomorrow dot eye oh"—got started in Boston in 2016, after Elkabetz and his cofounders had served in Israel's armed forces and come to the United States to earn MBAs. The first time I spoke with Elkabetz, in 2019, the company was still known by its original name, ClimaCell. During his eleven years in the Israeli air force, Elkabetz told me, he'd had plenty of chances to see the importance of accurate meteorology.

Elkabetz argues that companies like Tomorrow.io are needed to disrupt the balance in the weather enterprise. "Weather-forecasting technology has been led by government agencies, period. How many domains in the world do you know that are still technologically led by governments? Imagine if we had to rely for COVID-19 vaccine development on government agencies? We would probably still be sitting at home." Elkabetz adds that he doesn't believe there isn't a role for government, only that private companies can drive innovation. "I will use the analogy of SpaceX," he tells me. "Nobody thinks that because we now have SpaceX, we don't need NASA."

Tomorrow.io's technology ranges across multiple domains, each of which would be venturesome enough on its own. On one front, the company taps some ingenious sources to collect hyperlocal observations. Wireless signals, which have become practically ubiquitous in modern society thanks to widespread use of cell phones, microwave relays, satellite signals, and more, can reveal information about atmospheric conditions. Because moisture in the air can subtly affect the passage of wireless radiation, with careful monitoring, observers can infer the state of the atmosphere from those changes. Though these measurements may not be up to the standards of dedicated weather instruments, Tomorrow.io says the quantity more than makes up for the quality. Even though any given measurement point might lack precision accuracy, the sheer numbers permit assembly of an accurate picture.

The company says it draws on other sources, including the Internet of Things, or IoT. The utility of IoT stems from the enormous number of everyday devices connected in some way to the internet. Modern automobiles represent a fleet of roving sensors, equipped with the ability to detect temperature (often displayed on a dashboard readout) and precipitation (used for automatic windshield wipers). Cars with self-driving features like lane-keeping have even more sophisticated sensors. Some smartphones measure barometric pressure to aid in determining altitude for mapping purposes. Traffic-camera images and other web video feeds can indicate sunny or rainy conditions when processed with artificial-intelligence techniques.

Next, Tomorrow.io feeds this data along with traditional radar and satellite observations into its own numerical weather-prediction model, CBAM, which it says has been designed for hyperlocal flexibility. Depending on a client's or industry's need, CBAM can run at a coarser or finer resolution—down to tens of meters, if desired—to focus on a specific area, and it can be updated as often as every few minutes, Tomorrow.io says. CBAM illustrates the advances made in high-speed computing. No longer is the supercomputing power needed to run weather-prediction models limited to governments.

Yet even more remarkable is Tomorrow.io's plan to operate its own satellites. With support from a US Air Force contract to supply weather data, in 2021 the company announced its intention to place a network of satellites in orbit. These satellites are intended to provide radar and microwave observations on precipitation and temperatures. Just as the lower costs and growing availability of supercomputer-level processing make it possible for the company to run its own weather-prediction model, new approaches to satellite design enable its space effort. Tomorrow.io will use a constellation of up to a few dozen small satellites to achieve global coverage. Often referred to as SmallSats or CubeSats, these miniature devices get the design time and cost of an individual satellite down to a fraction of that of a traditional craft. (One notable example is the Starlink constellation launched by Elon Musk's SpaceX to offer satellite internet service around the world. SpaceX has launched thousands of the Starlink craft.) Contrast that with NASA and NOAA's workhorse weather-satellite program, which uses just a few large craft to cover the entire Western Hemisphere. Their capabilities are much greater, but so are their price tags. Tomorrow.io launched its first satellite in April 2023.

Why would a start-up go to the effort of designing its own sensor networks, computer models, and even a space program? Elkabetz says Tomorrow.io wants to be able to tailor its forecasting to each customer's needs. One early customer was JetBlue, which wanted precise forecasts targeted at key airports so it could better anticipate weather disruptions and work around them. In a case study, Tomorrow.io described a forecast for Boston's Logan International Airport during a February snowstorm. Most predictions called for snow from early morning until eleven a.m., prompting some airlines to delay or cancel flights. But when the hyperlocal Tomorrow.io forecast pinpointed the end of the snow at Logan at closer to eight a.m., JetBlue avoided cancellations and delays, according to the case study. (JetBlue went on to invest in Tomorrow.io.) Tomorrow.io has also worked with Uber, which wanted weather and forecast data to sharpen its arrival-time estimates, and the National Football League, which sought predictions about game-day conditions.

Elkabetz says all these technologies together allow Tomorrow.io to provide robust forecasting without some of the massive investments—such as ground radar stations and geosynchronous satellites—long associated with world-class weather services. "This is not only going to help businesses, it's actually going to help countries," he says. "With this business model, with these capabilities, we can become a meteorology agency as a service for many countries that do not have the luxury and the ability to invest half a billion dollars in ramping up radars, models, scientists to run it, a mini-NOAA."

I personally don't need to coordinate airport de-icing crews or worry about fertilizing crops, but I am pretty interested in the weather in my own backyard. And I don't just mean that metaphorically. My wife and I make frequent trips from New York City to our small family home upstate. The region is rural, and the terrain varies from flat meadows and farmland to steep hills and narrow valleys. When you walk, hike, or drive in this area, you see microclimates in action as winds shift around the hills and rain falls in spotty patches. A one-size-fits-all forecast can't always tell you what will happen when.

So I brought hyperlocal forecasting to my backyard in the form of a foot-tall white plastic device that looks like a cross between a flashlight and a back massager. It's my own personal weather station. It tracks the conditions right at my doorstep. It also lets me, in a small way, participate in the refining of hyperlocal forecasts because it feeds its observations first to the Wi-Fi network in my home and then on to the internet, adding data to a network of these personal weather stations.

My weather station is a Tempest. It has won praise from both weather hobbyists as well as some professional meteorologists for its simplicity, compact design, and ease of setup. The engineers have packed a surprising amount of technology into the Tempest. A light sensor on top keeps track of sunlight and UV radiation, charting the variations between sunny and cloudy periods. Other sensors measure temperature, barometric pressure, and humidity. One component detects the electromagnetic signature of lightning flashes and, based on the intensity, estimates whether the lightning is nearby or twenty-five miles away.

Another sensor, an ultrasonic anemometer, records wind speed and direction by detecting how the movement of outside air affects the patterns of ultrasonic waves generated inside the device. (I had to make sure to correctly line up an arrow denoting north when mounting the gadget.) And when falling raindrops strike the domed top of the Tempest, it senses vibrations and estimates the amount of precipitation from the size and frequency of the *tap-tap-tap*. A miniature solar panel powers the whole thing, including the wireless signal that sends the data to a small base station indoors and then to my home Wi-Fi network. The device has been constructed to eliminate moving parts, improving reliability.

Of course, meteorology geeks have long operated their own home weather stations, some relatively simple, others big and complex, sprouting spinning-cup anemometers and other sensors. The approach of the Tempest has some limitations; for instance, the vibration sensor for detecting rainfall isn't as precise as traditional precipitation monitors, where the falling liquid accumulates in a container or passes through a gauge. The Tempest sells for $339, while fancier stations can cost upwards of $1,000. A friend of mine operates a more elaborate station from Davis Instruments; he likes to tinker and connected the device to the internet by hooking it up to an inexpensive minicomputer called a Raspberry Pi. Ambient Weather, another company in the space, also makes some popular stations suitable for home use.

After I got my Tempest up and running, I quickly found myself addicted to checking the conditions at our upstate home, even when we weren't there. The data flows to a Tempest app on my iPhone, where I can see both live measurements (it's 54 degrees and sunny there as I write this) and prior readings and trends (no rain whatsoever for the past week). I get an iPhone notification whenever rain begins. One feature I've found surprisingly useful when I'm actually at the house is a lightning-detection alert. Because the Tempest can sense lightning strikes that are still a few dozen miles away, these alerts warn you when thunderstorms are on the way and it's time to wrap up grilling and get everyone inside.

In addition to those real-time observations, the Tempest app provides a forecast for my home's location, with near-term predictions displayed in an hour-by-hour format similar to other weather apps. But despite the appearance, this forecast is different. Tempest says it customizes the predictions for my home's location by starting with the big models from NOAA and the European Centre for Medium-Range Weather Forecasts, factoring in weather-station data to produce more granular regional forecasts, and then, finally, using artificial intelligence technology to produce a hyperlocal home forecast based on data from my Tempest station.[95] Though I haven't conducted a rigorous scientific evaluation of the predictions for my home's location, I have definitely found them more helpful than the generic forecasts.

"We're focusing on short-term forecasting," says Buck Lyons, the CEO and cofounder of WeatherFlow-Tempest, the company behind my home weather station and those hyperlocal forecasts. He says the business is focused on geographic scales between one and four kilometers, and in some areas even more fine-grained than that. "We're trying to deliver information to people where there are differences in the weather at that scale, and where they will really benefit from knowing what it's doing now, what it's going to do later today, and what it's going to do tomorrow." Lyons describes himself as an entrepreneur with an interest in science but an aptitude for business. "Working on weather and weather modeling really scratches that itch of being an amateur scientist, so I just love that," he says.

Though the consumer-friendly Tempest station has brought wide exposure to WeatherFlow-Tempest, it's only a slice of the business. The larger enterprise is taking the data from all the stations out there, refining hyperlocal predictions, and then translating the forecasts into information that consumers and businesses can act on. On the consumer side, for instance, WeatherFlow-Tempest envisions weather data as an integral factor for smart homes. Tempest owners can already direct a number of smart gadgets based on outdoor conditions. When the sunlight sensor detects a brightening sky, it can command motorized blinds to lower. Winds whipping up? The system can tell Wi-Fi-enabled lawn sprinklers to temporarily shut down lest the gusts send the streams astray, wasting the water. The possibilities will expand as smart-home devices proliferate. For businesses, meanwhile, WeatherFlow-Tempest promises hyperlocal forecasts that can grow in accuracy as more weather stations get deployed.

When it comes to the public/private tensions in next-generation weather forecasting, Lyons says government efforts serve as the foundation but companies like his can develop approaches that don't make sense as a public service. "I get frustrated with people who say, 'Oh, the government should do all of this, and then it should be available to everyone,'" he tells me. Taxpayers, he argues, shouldn't be on the hook for bespoke forecasts aimed at insurance companies or even hard-core home gardeners. "But

I'm also afraid of people who talk about how the private sector is the solution to everything. We still need big satellites and we need the government to do a lot of weather modeling." Lyons says it takes a balance, and he believes the weather enterprise is navigating the equilibrium successfully with companies like his innovating new services that leverage the massive public infrastructure. "I'm very pleased to see that we seem to be marching in the right direction," he says.

There's little question that more observations taken at more locations can improve hyperlocal forecasts. But multiply the feed from my home weather station times many others across the United States and you get a lot of data. As hyperlocal forecasting evolves, the challenges of figuring out how to gather that data, apply it effectively, and ensure its quality will only grow. To understand just how complicated a problem this is, it's worth looking at a weather experiment that echoes a pioneering project from the days when people were first discovering the power of the internet.

Back in 1999, a group of scientists offered home-computer users everywhere a remarkable opportunity—to participate, if only in a small way, in the search for alien civilizations. Their project was called SETI@home, and it was launched out of the University of California, Berkeley. Researchers in SETI (the Search for Extraterrestrial Intelligence) had huge amounts of data from radio telescopes pointed at the stars. But sifting through all that data for evidence of signals that could be ascribed to some possible intelligence rather than just random radio-frequency radiation required computer analysis.

The SETI@home scientists had multiple goals. One was simply to get members of the public excited about science in general and SETI in particular. Another was to exploit a massive untapped source of computing power; personal computers everywhere were increasingly connected to the internet but sat idle for hours at a time while their owners were away or asleep. So they created a software package that anyone could download and install. It functioned like a screen saver. When a computer sat unused for a specified period, the software would light up, download data files, run the mathematical analyses, upload the results, then do it all over again. SETI@home proved immensely popular, attracting more than five million participants before the program went into "hibernation," as its leaders termed it, in 2020.[96] Though we're still waiting to find alien signals, SETI@home demonstrated the potential for crowdsourcing scientific work. Similar projects have tackled protein-structure analysis to aid biomedical research and simulation of heart-cell behavior for cardiac-health insights.[97]

In hopes of gathering more information related to weather, the European Space Agency turned to crowdsourcing to utilize a growing worldwide network of computer-controlled sensors: the modern smartphone. Just as those early 2000s pioneers could download the program from SETI@home to put their PCs to work, smartphone owners can install an app to gather atmospheric data. The program is called CAMALIOT (for "application of machine learning technology for GNSS IoT data fusion," a convoluted acronym even in the jargon-laden world of space and atmospheric science).

"We now have these devices that are everywhere, that everyone has in their pockets," says Vicente Navarro, a satellite-navigation expert and project manager at the European Space Agency. CAMALIOT works by gathering data from the Global Positioning System receivers in smartphones. Most smartphone users are familiar with the basics: the GPS chip can locate your phone's position by receiving signals from satellites in orbit around the Earth. How that actually operates is more complicated. It depends on extremely precise clocks. The satellites continuously broadcast signals that include two key pieces of information: the exact time the signal was sent and the position of the satellite at that time.[98] Inside your phone, the GPS compares the time on its own clock with the time on the signals received.

Because the radio signals travel at a known speed, the system can look at the slight variations in the time codes to determine the distance from each satellite. As long as your phone receives signals from multiple satellites, it can triangulate its location based on the position of each satellite. It's worth noting that even though we use the term *GPS* to refer to navigation-satellite technology in general, it's actually the name of one specific set of satellites, the network created and owned by the US government. There are others, such as the European Union's Galileo satellite network, Russia's GLONASS, and China's BeiDou. The true generic term is GNSS, for global navigation satellite system. Many smartphones now have GNSS chips that can pick up signals from multiple networks.

So far, all that describes only how your phone figures out its location. To understand how CAMALIOT works, you need to know that those satellite radio signals are shaped by their travel from orbit and through the Earth's atmosphere to your phone. "The water vapor in the air is actually what we can measure," says Benedikt Soja, project manager for CAMALIOT and a professor of space geodesy at ETH Zurich, a prestigious Swiss university focused on science and technology.[99] He explains that moisture affects the radio signal on its way through the atmosphere. "It slows it down to a certain extent. Also it bends, so there's some refraction going on," Soja explains. "This is quite interesting, because we can very precisely measure the water vapor between the satellite and the receiver."

This behavior underlies an established source of atmospheric data referred to as radio occultation. CAMALIOT applies this science to the satellite signals received by mobile phones to make map apps work. When the CAMALIOT app is installed on a phone, it can access the raw signal data from the phone's GNSS receiver, store the data, and upload it to the CAMALIOT servers, where it is analyzed to derive the moisture content.[100] If you're interested and have an Android phone, you can download the CAMALIOT app from the Google Play store. If you have an iPhone, you're out of luck—the Apple system doesn't allow apps to access the raw GPS signal data, Soja explained to me. Due to privacy concerns, no personal information gets stored with the GNSS data.

We are talking about a lot of data here, and that's part of what distinguishes the CAMALIOT research from more established uses of radio-occultation information. After less than a year, according to program officials, more than eleven thousand CAMALIOT users in 142 countries contributed over a hundred billion individual measurements. What Soja and his colleagues really want to figure out is how to turn these massive datasets—on the scale expected from tapping crowdsourcing and the Internet of Things—into something usable. That involves not only collecting the information but sifting through it and tossing out poor-quality data. "Recently, with developments in AI and machine learning, this has actually become feasible," Soja says. "In the past, if you were to go through the data manually, you would not have any chance to succeed."

So far, CAMALIOT remains experimental. There continues to be a tension between traditional measurements collected from purpose-built, highly calibrated sensors and the gusher of data—often highly imperfect—from the Internet of Things. But the ability to gather observations from practically anywhere promises finely detailed information that can improve hyperlocal forecasts. And the success of CAMALIOT in attracting users suggests a bigger role for crowdsourced weather data in the future. "It was a huge surprise to see the capacity to motivate people to contribute as citizen scientists," Navarro says. "People were really looking for a way to contribute."

Hyperlocal forecasting looks set to drive some of the most interesting applications for meteorology in the years ahead. Though it remains to be seen just how accurate these predictions can get, demand for weather intelligence keeps expanding as more businesses prepare for the impact of extreme events from climate change and as a way to create everyday efficiencies. Whether it's an airline using Tomorrow.io's pinpoint guidance to schedule de-icing operations or a farmer like Andrew Nelson keeping tabs on freeze conditions across different parts of his land, the applications for hyperlocal forecasting are already clear. And the work going into developing these kinds of solutions propels disruptive technologies, like Internet of Things observations, that can bring better forecasting to people around the world in regions that lack robust weather infrastructure.

Of all the realms of forecasting I've learned about in my reporting, hyperlocal displays some of the greatest potential for innovation. It also affirms the role for businesses to advance forecasting even as the other legs of the weather enterprise tripod—government and academia—are critical to progress in everything from basic science to public safety. Examples and experiments popped up everywhere as I met and talked with weather experts.

At Waymo, the self-driving car technology company from Google's parent company, Alphabet, engineers are designing automobiles to be weather-smart. Waymo cars use an array of onboard sensors, from LIDAR to cameras, to assess the environment. The company drew on all of this data to create a high-resolution map of fog in the San Francisco area.[101] Waymo now describes its self-driving cars—which are already on the road in San Francisco and Phoenix, providing rides that users can hail through an app—as "mobile weather stations." Observations gathered by the Waymo fleet help the company determine at a hyperlocal level where weather conditions are improving or deteriorating, the company says.

In 2022, a Swiss company called Meteomatics began offering forecasts based on a high-resolution numerical weather-prediction model covering all of Europe. They dubbed the model Euro1k because it localizes its predictions down to one kilometer. Meteomatics supplements the traditional observations that feed into a computer model with readings taken by its Meteodrones, specially designed drones that soar up to nearly twenty thousand feet to sample the atmosphere. The company sells forecasts to shipping, energy, agriculture, and other industries. And Understory, based in Madison, Wisconsin, refines forecasts with its own weather station, the Dot. One focus for Understory is assessing weather and climate risk for insurance purposes, such as policies for auto dealerships, which can suffer substantial losses if a storm dumps hail on uncovered cars.

The list goes on. Many of these companies focus on business needs, but consumers also have access to better hyperlocal forecasting now even if they don't have weather stations like the one silently gathering observations at my upstate New York house. Generally consumers care most about temperature and whether it's likely to rain (or, in colder periods, snow). A smartphone provides easy access to basic hourly forecasts from a variety of apps. Apple says its own weather app, which comes on every iPhone, incorporates Dark Sky, a popular hyperlocal app that Apple acquired in 2020. Dark Sky began more than a decade ago with a crowdfunded Kickstarter project from developers Jack Turner and Adam Grossman, who had the idea to take data from the National Weather Service network of radar stations and process the information to look for precipitation and analyze where that precipitation would be in thirty minutes or an hour.[102] Most of the big names, including AccuWeather and the Weather Channel app powered by the Weather Company, offer their own tailored forecasts. If you're curious, you can visit Forecast Advisor (www.forecastadvisor

.com), a site that compares predictions from a number of these sources with the conditions that later occurred. Type in your zip code, and the site will show you accuracy data for your area.

For my part, when I get curious about rain headed to my precise location, I often skip the app forecasts and turn to the raw data from a smartphone app called RadarScope. I learned about RadarScope from professional storm-chasers; they say it's a dependable, easy-to-use, and powerful interface for the NWS radar-station network. With a little experience, you can easily see where it's raining nearby and whether the precipitation is headed your way. The NEXRAD Doppler weather radars operated by the Weather Service provide incredible detail on what's happening. Storm-chasers use that radar data to monitor the formation of supercell thunderstorms and to look for hook echoes, curved patterns on the radar that can indicate a tornado. RadarScope also helpfully displays the map boundaries of any watch or warning areas for all Weather Service–issued alerts.

My own usage has been more mundane, though valuable nonetheless. I was able to impress my family at a New York Mets game by noting the incoming rain on RadarScope and suggesting we retreat to the stadium's interior and find indoor seats in a lounge area. We did, and a few minutes later, a heavy shower began and drove thousands of other Citi Field fans inside, dripping wet and searching for spots to wait out the rain delay. We relaxed in our seats, warm and dry.

Back in Virginia Beach, a DroneUp crew prepares for a delivery mission. The flight team operates from a corner of a Walmart parking lot; they gather orders placed online by customers and get them loaded onto the quadcopter. A pop-up canopy, similar to the kind vendors use to shield their wares at an outdoor market, shelters a lightweight table serving as a control station. There's a desktop computer, two big screens, and several rugged-looking black cases housing communications gear. In contrast with all this high tech, there's an inexpensive folding chair for the operator, the kind my grandparents referred to as a card-table chair. Within easy reach sits a three-ring binder labeled *RPIC Checklist & Comms*. (RPIC stands for "remote pilot in command," the team member with overall responsibility for ensuring a safe flight.) The drone itself sits several feet from the tent. Orange parking cones and plastic barriers mark off the entire area.

"Stand by for mission brief," says Johnathan Wheeler, speaking into a walkie-talkie. "Package weight three pounds fourteen ounces. Current weather conditions, we're looking at fifty-four degrees, twelve-mile-an-hour winds, ten statute miles of visibility, and no cloud ceiling." Wheeler is the RPIC for this delivery flight. Also on the team is a safety pilot backstopping the RPIC and a visual observer tasked with monitoring the sky and the ground for any obstacles or dangers. Wheeler tells me he first became interested in UAVs back in college, where he studied homeland security and emergency preparedness. He's certified by the Federal Aviation Administration as a remote pilot. To get this flight off the ground, he works methodically through a checklist. The other team members read back elements of the flight plan and report their readiness.

As always in aviation, weather conditions matter. Amateur pilots and airline crews alike check on the weather for every flight by consulting reports known as METARs and TAFs.[103] Wheeler and his DroneUp colleagues do the same to get ready for each flight. A METAR gives pilots the current conditions at an airport, including temperature, wind direction and speed, visibility, and more. METAR data typically comes from one of the automated weather stations located at airports and operated by the National Weather Service with the Federal Aviation Administration and the Department of Defense. A TAF provides similar information but in the form of predictions rather than summaries of current conditions. Forecasters in local NWS offices issue TAFs for airports in their regions, looking ahead twenty-four to thirty hours.

The weather-information needs of DroneUp and other UAV-delivery start-ups are unusual and underserved. Wind speeds represent a critical factor. Safety requires keeping drones on the ground during excessive winds. Even when the winds are acceptable from a safety perspective, they play a big role in the efficiency of UAV operations. Sending a drone sailing into headwinds means working those spinning rotors harder to keep up forward speed, draining the craft's batteries quicker. Shifting winds can make hovering more complicated. Temperatures affect efficiency, as high-tech batteries don't perform well when they're too hot or cold. Precipitation can hamper visibility and affect the operation of onboard sensors.

John Vernon, DroneUp's chief technology officer, tells me wind handling presents a particular challenge because of the lower altitudes where drones spend most of their time. Generally people care about the winds on the ground and high aloft, the realm of commercial airliners. And when airplanes take off and land, they typically do so at airports with their own weather stations. "There's a huge opportunity that needs to be filled in terms of actually creating good solid wind data at these lower altitudes," Vernon says.

Though DroneUp's flights are fully operational—as I write this, you can visit the company's website, type in your address, and, if you're near a participating Walmart, order up your flying delivery—the availability is relatively limited. For most shoppers at the moment, the service remains something of a novelty. Scaling this new industry up will require infrastructure that makes the flights routine, with weather forecasting and data integrated into the systems. DroneUp partnered with Iris Automation, a company providing safety technology for uncrewed aviation. (Iris has since been acquired by uAvionix, another aviation-safety company.)

Iris developed collision-avoidance systems critical for drones making longer flights, called BVLOS (for "beyond visual line of sight") missions. Cameras on the ground and on the UAVs themselves scan the skies, and machine-learning software looks for any image that might represent another aircraft or some sort of obstacle. To incorporate weather into its products, Iris worked with TruWeather Solutions. Don Berchoff, the meteorologist and retired air force colonel who cofounded TruWeather, says his military experience taught him the need to bottom-line the weather for users. "Everything you do in the air force is about the mission and making sure that mission happens," he says. "So everyone that I dealt with, they didn't want to hear about the science. They just wanted to know if they could fly."

TruWeather draws on a number of data sources to create hyperlocal forecasts (the company prefers the term *microweather*) to support both advance planning and flight-time operations. For instance, Berchoff explains, if drone operators learn a weather system will move through the area from west to east the next day, they can plan flights to focus on the east side earlier in the day, before the winds hit, and the west side later in the day, when the winds have already moved through. Ultimately, though, the weather becomes just one more factor in planning a safe flight. As Jon Damush, the CEO of Iris who went on to become CEO of uAvionix, put it: "We're not building a business to be a weather company. We're building a business to be an AI-for-aviation company."

For now, though, the human factor looms large. As Walmart shoppers park their cars or exit the store carrying their purchases, the DroneUp team meticulously reviews their flight plan and calls up weather conditions from nearby airfields. "Parameters confirmed," Wheeler announces into the radio. "Flight deck, prepare for takeoff." The propellers on the drone spin up, and the craft, with a package tucked beneath, rises into a clear blue sky.

5 Extreme Heat

HOW TO THWART A SILENT KILLER

It is 5:43 a.m. on a mid-July weekday. The Northern Hemisphere summer promises more than fourteen hours of daylight, and the sun is already brightening the sky to the east. I'm standing on Broad Street in the heart of downtown Richmond, Virginia, a boutique hotel lobby to one side and a trendy coffee shop across the street. You don't need to venture far to appreciate Richmond's charms. The James River runs through the city, its banks lined with parkland that offers bicycle trails and walking paths. But you can't wander far without bumping into reminders of Richmond's complicated history. A few blocks from the columns of the state capitol building is the Shockoe Bottom neighborhood, where slave auctions were once held. It's a short walk to the Library of Virginia, home to a repository of slave-trade records that draws Black Americans hoping to fill in the blanks of their genealogies.[104] These days Richmond, home to the University of Richmond and Virginia Commonwealth University, has a cosmopolitan character. But the legacies of inequality here are inescapable. And, though most people rarely stop to think about it, those inequalities continue to play out in a map of the city's weather. In Richmond, some of the hottest parts of the city—the places considered urban heat islands—overlap with the geography of historical housing discrimination and racism.

At this early hour, Broad Street is just waking up. Even before six a.m., the air feels oppressive, the stillness and touch of heat a suggestion of a sweltering day to come. Today I am here to learn the intricate reality of a seemingly simple question: How hot is it? A quick check of my iPhone tells me that it's 72 degrees Fahrenheit, but that's too easy an answer.

After a few minutes, an SUV pulls up, my ride for the morning. It's a Chevy Trailblazer, unremarkable except for a few odd contraptions projecting from the front passenger door. Pinned between the top of the window and the door frame is a mount for a narrow white tube that juts up above the vehicle's roof. The top of the tube intersects a wider, shorter cylinder in a T; visible inside are some electronics and wires. Next to this mini-periscope is a second device, a more compact white plastic gadget shaped vaguely like one of the ghosts from Pac-Man, an upside-down U with chevrons at the bottom.

We are setting off to take Richmond's temperature, minute by minute, block by block. We will follow a 10.2-mile course along main roads and side streets, mapped out to capture specific varieties of terrain. When we embark on this drive, other volunteers will set off on additional routes across the city, all piloting cars that have sprouted white plastic periscopes. This intense one-day effort is called an urban heat-island mapping campaign. It's a relatively new undertaking but one that is spreading fast across the country. In the summer of 2021, when I joined the volunteers in Richmond, others were fanning out to map temperatures in the Bronx and Manhattan, Charleston, South Carolina, and multiple other cities around the nation. But Richmond has played a special part in the efforts to collect observations and understand the meaning and implications of the data.

"We're going to start in this neighborhood, this is called Bellemeade, it's historically an African-American neighborhood," explains the driver piloting our probe vehicle. He is Parker Agelasto, executive director of the Capital Region Land Conservancy, a nonprofit trust that acquires properties to preserve them and protect land and water resources. "You'll see Bellemeade Park off to the right. There's been a huge amount of attention to creating that park and making greenways so that students can safely access their school, which is next to the park." Agelasto turns to his copilot, Peter Braun, a young colleague at the land conservancy, for directions on the route we're supposed to follow. Braun tells him to keep going, then turn right on US 1 North.

As we begin our mission, the hottest temperatures of the day are still about nine hours away. This run has been scheduled for six a.m. so observers can take the city's temperature at the coolest time of day, after the Earth has had the entire night to give up heat it absorbed from the sun's rays the previous day. But in a small way, the complexity of that seemingly simple question—*How hot is it?*—has already become apparent. One answer stares at us from the dashboard of the Chevy. Like many cars these days, the instrument panel displays a measurement of outside temperature, typically from an ambient temperature sensor located behind the grille where outside air enters the engine compartment. Not necessarily a highly calibrated instrument, but useful. The number on the dashboard right now is 74 degrees.

I check my phone and get another answer: 72 degrees. This is the official temperature for Richmond at the moment, reported by the National Weather Service. This reading comes from one of the Automated Surface Observing System (ASOS) stations that provide measurements of key conditions such as temperature, wind, precipitation, and humidity. The ASOS program is operated jointly by the Weather Service, the Federal Aviation Administration, and the Department of Defense and boasts more than nine hundred stations, located primarily at airports.[105]

Here today in Richmond, the most interesting answer to *How hot is it?* won't be found on the Chevy's dash or on my phone—it's in the data being captured by the precision instruments fixed to the front passenger door. The core temperature data from the periscope-style probe flows into a memory chip for later downloading, so we can't see the live readings. But there's also the AirBeam, that Pac-Man-ghost-shaped air-quality monitor. It has its own temperature sensor, and it's feeding live data to a tablet computer, giving us a peek into the current conditions. And right now the readout shows 78 degrees.

How hot is it? Three different sources give us three different answers. In this case, even though the number many people would rely on for a quick check of the weather (the one displayed by my iPhone's weather app) says it's 72 degrees, the most precise answer we have (from the high-tech AirBeam) tells us that it's 78 degrees. In other words, at this exact moment and location, the secret reality of the heat is that it's six degrees hotter than the reading most people have easy access to. And it's even more complicated than that. As our Chevy roams avenues and side streets, the temperature reading shifts. For the sixty minutes of this data run, as we wind our way across the carefully designed route, we're assembling a map of the hottest and coolest spots in Richmond. Later in the day, in the afternoon's high heat, another team will retrace this same route. And it will get driven again in the evening as the city begins to cool.

The information harvested tells us how the layout of Richmond was shaped by the past, how the city's residents experience heat in the present, and, perhaps, how weather forecasts and the answer to the question *How hot is it?* will help those most at risk from extreme heat in the future.

Heat is the great deceiver among weather dangers. It isn't acute; it's literally a matter of degrees. A 100-degree day looks pretty much like one that's 85. There isn't much to see. The skies don't darken in warning. The danger isn't instantaneous, like a gunshot. It's slow-moving and cumulative, like a poison. Heat doesn't offer much in the way of visuals. Hurricanes announce themselves with giant whorls on a satellite image and leave behind shattered buildings and flooded streets. Tornadoes trigger piercing alert sirens and provide clips of terrifying funnel clouds. Heat, however, just is. Apart from the occasional video of someone frying an egg on a sidewalk (once a reliable local TV news stunt, now fodder for TikTok), deadly heat doesn't look like much. It just quietly kills people.

Of all the perils described in this book, heat may be the most paradoxical. We are quite good at forecasting the temperature, but we have far to go when it comes to using those predictions to keep people safe. Our very skill at foreseeing dangerous heat waves makes some of the tragic outcomes all the more agonizing.

Heat kills more people in the United States than any other type of extreme weather. Each year the National Weather Service puts out a report summarizing weather-related deaths and injuries by category. In 2020, for example, the Weather Service reported 350 heat fatalities for the United States, by far the deadliest category. Tornadoes, by contrast, killed 76.[106] It's critical to understand that such tallies are widely presumed to undercount heat fatalities because of the different ways high temperatures affect the body. Acute heatstroke can kill swiftly, but heat can kill indirectly as well. Heat stresses the body and can lead to fatal events, such as heart attacks or kidney failure, in those with certain medical conditions, and these victims may not succumb for hours, days, or even weeks.

To better understand the possibilities for progress, I spoke—of course—with meteorologists. But I also talked with doctors, public-health officials, emergency responders, and community leaders. They made it clear that a good forecast is just the start. Scratch that—the start must come well before the forecast. Keeping people safe from heat means getting ready far in advance. Some preparations can be completed in days, but others require months or years. Just as coastal areas threatened by flooding must construct infrastructure like seawalls, communities that want to fight heat need everything from buildings cooled by green roofs or reflective coatings and rugged power grids able to handle air-conditioning demands to cultural and educational efforts that encourage residents to react to heat dangers with the same alacrity they would in a tornado or hurricane emergency. All these efforts necessitate helping people to understand the threat, making sure organizations and cities are ready to respond, and increasing society's resilience in the face of heat. And they demand that we understand what individuals can and cannot do to protect themselves and how the most vulnerable, from homeless people to the elderly to low-wage outdoor laborers, have the fewest options when the temperature climbs.

Heat has long been the least appreciated of the weather dangers relative to the number of victims it claims. But with each year, the problem grows more urgent. Climate change drives both the frequency and the intensity of heat waves.[107] It also pushes hotter weather into cities and regions that are less accustomed to these temperatures and are therefore less prepared. In the years to come, the forecasts will call for heat more often, and we can rely on those forecasts days in advance. In fact, the ability to predict a heat wave keeps getting better. What matters is how we respond.

Heat is very bad, but what can we do about it? One surprising answer is an object measuring about three feet wide by two and a half feet tall and composed mainly of plain unfinished wood boards. The fancy name for this object is a *heat-respite area*. If you saw one in real life, you would probably think it looks like a sidewalk planter with a built-in bench to sit on.

You'd be correct. The heat-respite area I'm describing here is a wooden bench integrated with a planter behind the seat. It can often be found sporting a patio-style umbrella sprouting from a bracket on one corner. The story of how this object came to be and how it's helping one Philadelphia neighborhood respond in a tangible way to the dangers of heat tells a great deal about the challenges of this weather threat. At one level, the bench's function couldn't be simpler: It gives residents a place to sit outdoors in the shade on a broiling sunny day. But it plays an outsize role in encouraging the neighborhood's residents to take hot weather seriously. It's a story that encompasses the tension between high-tech forecasting and low-tech solutions and the payoffs possible when we confront weather dangers at a community level. And it's a tale of finding small ways of engaging people in larger, crucial conversations about climate and weather resilience.

This story begins in Hunting Park, a neighborhood about five miles north of Philadelphia's Center City. It's easy enough to find: Start at City Hall and follow Broad Street north, beyond Temple University and past the intersection where Germantown Avenue slices on an angle across Broad (home to the cheesesteak spot Max's Steaks). Most of the residents of the Hunting Park neighborhood are Hispanic or Black, a result of gentrification in areas closer to downtown that pushed minority residents north. The neighborhood takes its name from the nearby city park, bounded by North Ninth Street, West Lycoming, and Old York Road.

One of the clichés of heat warnings is an announcement you've probably heard, particularly if you live near an urban center: "Cooling centers will be open today." If you're privileged enough to have air-conditioning, you may have little more than a vague idea of what a "cooling center" is, but your vague notion is probably enough. A cooling center is simply a public space that's air-conditioned. It might be a community or recreation center, a public library, or even a lobby or an empty room at a government office, such as a city hall or police station. The idea is to make sure that people who don't have access to air-conditioning—particularly vulnerable people—have a place to go when temperatures reach dangerous levels. Some officials I talked with told me that it isn't always easy to get people to take advantage of cooling centers for reasons that range from a reluctance to leave their homes to a sense of stigma associated with it, akin to needing a homeless shelter. (In fact, some of the most at-risk individuals are homeless people.) Adding some type of activity or event, such as a book reading at the library or games at a rec center, can make a visit to a cooling center more attractive. As with so many aspects of heat dangers, economic and social disparities shape the problem. Wealthier people can afford to buy air conditioners and, just as important, won't balk at the electric bill for running them. According to the federal Energy Information Administration, the annual tab for home air-conditioning in a warmer region can easily top four hundred dollars. So, regardless of their flaws, cooling centers have been an important tool in heat safety.[108]

Or they were until the arrival of the COVID-19 pandemic in 2020. As lockdowns spread that spring, Franco Montalto, a Drexel University professor of civil, architectural, and environmental engineering, asked himself: What are people going to do when summer comes to Philadelphia? Montalto had seen seasonal forecasts projecting above-average heat for the region. "Philadelphia, like many cities, was closing down libraries, adult centers. Cooling centers, they were all closed. So, there was this question in our mind. Well, we're going to have to have something," Montalto says. "I started looking at how people in different parts of the world were addressing extreme heat in outdoor contexts and found a lot of interesting examples, from Japan to Europe to other places. And they fit into these three categories of shade structures, greenery, and pavement wetting."

Shade structures and greenery are easy enough to understand as ways to beat the heat. Pavement wetting is more unusual. In the city of Nice, in France, officials have deployed two different approaches, according to Europe's Covenant of Mayors for Climate and Energy, a coalition of local governments.[109] One approach feeds water beneath a road or sidewalk composed partly of permeable materials that allow the liquid to percolate up, eventually reaching the surface, which it cools by evaporating. The other technique used in Nice involves spraying water onto roadways, again tapping the cooling power of evaporation. High-tech implementations aside, pavement wetting dates back centuries in

Japan, where it's known as *uchimizu* ("sprinkling water") and done with buckets and ladles.

Montalto took these ideas to Saleem Chapman, who in 2020 was appointed Philadelphia's first chief resilience officer, a position devoted to addressing climate impacts that's being instituted by more cities as they size up current and future implications of climate change. Miami, for instance, now has a chief heat officer tasked with developing responses to rising temperatures and the inequities that cause some Miamians to be disproportionately affected. No matter the title, it is the job of Chapman and counterparts across the country to find ways to avoid, mitigate, and otherwise respond to trouble. On a longer timescale, that means dealing with problems like rising sea levels that can turn once-sought coastal properties into flood zones. But in the shorter term, a resilience officer's job has a lot to do with the weather, whether that's assisting recovery after a hurricane or protecting people in a heat wave.

Chapman was intrigued by Montalto's ideas and had a suggestion: Consider trying out these ideas in Hunting Park. The neighborhood was considered a classic urban heat island. According to city figures, buildings, roads, and other paved surfaces covered more than 75 percent of Hunting Park but only 52 percent of Philadelphia overall. Tree cover for Hunting Park was 9 percent; it was 48 percent for a leafy neighborhood like wealthy Chestnut Hill. In 2019, Philadelphia mayor Jim Kenney made Hunting Park the focal point of the first community heat-relief plan in the city's history, dubbed "Beat the Heat Hunting Park." This was a comprehensive plan built on a year of research with extensive community involvement.[110] Among the initiatives outlined were plans to plant more trees and a commitment to factor heat vulnerability into city planning decisions, such as where to place bus routes so that residents could avoid excessively hot walks and extending hours at selected public pools. Kenney and other city officials rolled out this Beat the Heat plan with a splashy Hunting Park event, highlighting the neighborhood's vulnerability due to the relative lack of tree cover and stressing the health dangers.

As it happened, while Montalto was fretting about closed cooling centers, the William Penn Foundation—a Philadelphia-centric philanthropic organization that frequently supports environmental programs—was looking for projects to help offset impacts from the pandemic. Montalto figured that an effort to respond to the difficulty of running cooling centers could be a good fit. He secured a grant to fund a pilot of his idea on one street: the 4400 block of North Marshall Street.

The next step was to involve the community itself through participatory design, an approach that gathers input from all stakeholders to make sure the final outcome meets their needs. In Philadelphia, block captains are unpaid community leaders who receive recognition and support from the city government. Montalto met with a few block captains and developed an initial strategy that included distributing sprinklers for pavement wetting and providing shade structures.

"We started just by looking at the opportunities on the block," Montalto recalls. "And in fact, initially the inspiration was southern Spain, where they hang canvas from one side of the street to the other in strips, and it creates a cooling situation." The Hunting Park working group recognized that affixing things to people's homes might be complicated, so they concluded the shade provider should be a standalone structure. Next, they went through all the possible obstacles, making sure the work wouldn't run afoul of any city regulations. They wanted an umbrella for shade but they also wanted to make sure a windy day wouldn't turn the parasol into a missile, so they brought on a structural engineer. The design evolved into a compact, easy-to-build wooden structure with space on top to hold soil and plants and a flat bench to serve as seating. The planter/benches were born. They also devised a process for recording temperatures so they could determine whether the sprinklers and shade structures were indeed providing a cooling effect.

The pilot was a success. Residents on the test block accepted the presence of the structures on their street and turned to them for shade on days when the heat indoors was stifling. In 2021, with the pandemic still limiting residents' movements—particularly at-risk elderly Philadelphians—Montalto and his teammates decided to expand their work. This time around, they secured funding and set a goal of one hundred planters across five additional blocks. Crucially, they doubled down on the community nature of the product. "We recruited, trained, and employed ten workers for six weeks in carpentry and horticulture-related skills," Montalto says. They also hired local residents as "civic scientists" to collect and track data. By this point, Montalto says, "We had a product, we made some tweaks to it, but it was a product that we were able to produce rapidly and efficiently and make a hundred of them."

A key ingredient in making all this happen was a community organization called Esperanza that provided a trusted link with residents. Esperanza—the word means "hope" in Spanish—fills an enormous role in Hunting Park. Calling it a community organization doesn't come close to representing its many initiatives. Founded in 1986 by Reverend Luis Cortés Jr. with support from the Hispanic Clergy of Philadelphia, Esperanza has grown into a neighborhood anchor with more than 450 employees. It provides everything from housing counseling for people trying to navigate rentals and home purchases to economic development initiatives that support small local businesses and fund affordable housing. Esperanza operates a K-through-12 charter school and a community college, both focused on the heavily Hispanic and Black population in the area.

Jodi Reynhout, who was the senior vice president of community engagement at Esperanza, told me that her organization was already keenly aware of Hunting Park's vulnerability to heat and of the environmental-justice issues involved. It had been working for years on a number of fronts, including tree planting and gardening initiatives. Esperanza engaged in community outreach and education around heat and health, giving out heat-safety kits with water bottles and electrolyte tablets and distributing donated fans and air conditioners. It also worked on cooling centers, organizing with churches, libraries, and other locations with air-conditioning to set up the kinds of activity agendas that could help overcome people's reluctance to make use of them.

Then in 2020, Esperanza ran into the same problem that had so concerned Montalto. "Because of COVID, we couldn't do cooling centers," Reynhout recalls. "We had this whole idea, we had sites lined up, and then everything shut down. We said, Okay, what can we do instead? We had to pivot." And that's when Montalto showed up with his concept and funding, in need of inroads into the community. Esperanza helped loop in the block captains and other resident voices vital to engagement.

Ultimately, the deeply local nature of this project drove its success; a professor at a local university, the city's sustainability office, a Philadelphia-centric grant maker, a well-known local community center, and residents themselves came together. All this combined to take what could have been a fairly ordinary case of installing a neighborhood amenity and elevated it into something bigger. The core idea was to provide shady places for residents to escape their hot indoor rooms and spend time outside. But the collaboration led to much more. The community surveys and participatory design discussions got residents talking about heat. Hiring local workers to create the planters kept the project community-centric and spurred more conversations. And once the planters were installed, they became a visible symbol, stimulating even more discussion. It all coalesced into a powerful catalyst for getting residents of one of Philadelphia's hottest neighborhoods to focus on the health dangers from extreme heat.

When I visited Hunting Park and arrived at North Marshall Street, lined on both sides with row house–style homes, I could see the planter/benches dotting the sidewalks. One stood in front of the home of Priscilla Johnson. She is a mother of four, a grandmother of five, and the block captain for her stretch of Marshall Street. Johnson has lived here for more than thirty years. Her house is immaculate. She says she became a block captain after noticing how messy the street had become after the previous block captain moved away. "As far as cleaning, the sweeping up, I can't stand to see all the trash just sitting there," she tells me. This type of community self-policing is one of the original functions of the block-captain program; the program office that oversees it is called the Philadelphia More Beautiful Committee. "It's a lot of work," Johnson says. "People have to know you and know your spirit. With everything we've been through on this block, they know me by now." Through her work with the Beat the Heat initiative, Johnson learned about the 2020 pilot program and volunteered—making her block of North Marshall the first in Hunting Park to become home to the planter/benches.

"People weren't actually coming out in the heat. It was just too overwhelming," Johnson says. That changed after the planters arrived. "I came outside more than ever, sitting on my bench. Other people came and sat outside and the kids loved it." Johnson says neighbors liked the way the planter by her home looked and would come to ask her about it. "As soon as everybody was seeing what they were, I was getting all kinds of 'Oh, I want a bench!' And I explained to them, it's not just about beautifying your house. It's about heat. And that was the real message behind all this."

Johnson told me her own awareness of heat dangers has grown thanks to the planter initiative and the broader Beat the Heat campaign. In particular, she's more conscious of the heat island that is Hunting Park. "All I knew was it was just hot," she says. "I'm thinking I had to just deal with the heat, not knowing that we're in an area where it's especially hot." Johnson says she's also more aware of the health risks, particularly given her diabetes and high blood pressure. But she says it's not something her own doctor discussed with her much. "I think people need to be a little bit more educated."

Not far away, on North Ninth Street, across from the actual park that gives the neighborhood its name, two of the planter/benches are stationed in front of a town house. This is the headquarters of a youth-focused nonprofit community group called As I Plant This Seed, founded and run by Ryan Harris. The town house serves as a combination of classroom, meeting area, and neighborhood clubhouse. Harris tells me about growing up nearby and graduating from Cheyney University of Pennsylvania, an institution west of Philadelphia that was founded in 1837, making it the first of the nation's HBCUs (historically Black colleges and universities). "This is where I grew up," Harris tells me. "It's a personal place for me. It was important for me to build here."

There's a lot going on at As I Plant This Seed. Harris provides a mentorship program for middle- and high-school students, Talk to Me, that enrolls seventy kids at a time in ten-week series of meetings. The meetings cover everything from civic responsibility to financial literacy, but they also give kids a place to come together. "It's just a lot of things we teach that they're not getting at home or they're not getting at school," Harris says. "Kind of building a village for them, if you will." He's been doing this for a decade now, growing the organization year by year.

When Harris first saw one of the planter/benches, he says, he knew he wanted one outside his neighborhood center. It offers a spot for some of the kids attending his program to take a break in the shade. And just as on Marshall Street, it serves as a conversation starter. Harris sees heat and its disproportionate effect on his neighborhood as one more community issue that requires a response. Among other things, his group helps distribute air conditioners. "We're getting them to the people we know probably wouldn't be able to afford them or to the elderly that are literally not getting help or being cared about. Somebody's got to check on them," Harris says. "So we knock on doors."

For the kids he works with, the planter/benches start conversations about climate change and environmental justice. But Harris also strikes a realistic note about promoting these topics in a lower-income neighborhood where residents are worried about jobs and crime. "There's a lot going on in this community; people don't have time to think about climate change," he says. "That's not going to be the thing that'll kill you. A bullet will, way before climate change." Still, he says, when kids come to As I Plant This Seed on a day where the forecast calls for heat, they gravitate to the planters. "The first thing that the first couple of kids who come in do is open up the umbrella," Harris says. "Everybody knows it's about to get hot and we need some protection."

Is the shade of an umbrella outside as good as the cool breeze from an air conditioner? Maybe not. But the popularity of the planter/bench project in Hunting Park points to a number of factors that can encourage success in coping with extreme heat. It centered the conversation within the community from the very beginning. Along the way, this process reminded people of Hunting Park's disproportionate vulnerability to heat, respected residents' connection to their homes, and got them talking about heat health and safety. It suggested that people living in an urban heat island should take a Weather Service heat warning as seriously as, say, people in coastal Florida take hurricane alerts. And it gave individuals simple actions they could implement when a forecast said high heat was on the way: Open the umbrella, stay in the shade, take care of yourself, and check on your neighbors.

Why is heat so dangerous? It's considered the deadliest of severe-weather threats. Tornadoes and hurricanes get a lot of the press because they produce scenes of destruction and kill or injure people suddenly. It's easy to conjure images of television correspondents leaning into the wind while wearing logo-bedecked rain slickers or walking past the remnants of a shredded home. Heat is a stealthier danger. "The effects aren't instant," says Gregory Wellenius, professor of environmental health and director of the Center for Climate and Health at Boston University. "It's not like you spend two minutes in high heat and you're going to pass out."

The machine that is your body wants to operate at a certain temperature. Many of the body's functions are dedicated to making sure it stays at that temperature. Normally the body temperature for an adult is somewhere between 97 and 99 degrees Fahrenheit. As kids, we learn that 98.6 is normal when our parents check us for fever, but researchers have found that some people run warmer and others run cooler and that "normal" can vary at different times of the day. If you're old enough, you might remember having your temperature taken with a glass mercury thermometer with 98.6 highlighted on its etched markings—the legacy of a nineteenth-century German doctor named Carl Wunderlich who took the temperatures of thousands of people and concluded that the average normal reading was, you guessed it, 98.6. More recent research shows the modern normal temperature falls around 97.9.[111] Suffice it to say that your body is happiest when its internal temperature is somewhere between 97 and 99.

Under normal conditions, your body constantly sheds heat into the environment to maintain that temperature. Classically, there are four major mechanisms for the body to lose (or gain) heat. Radiation is a big one. Heat in the form of infrared waves can transfer energy, but this process depends on the temperature difference between your body and the environment. If the environment is hotter than your body, you can't effectively radiate heat away. Another mechanism is conduction, in which your body is in physical contact with something at a different temperature—think rolling a cold beverage can across your temple. Your body will transfer heat into the can, eventually warming it up. The third mechanism is convection: Since hot air rises, as your skin heats, the air immediately adjacent to it, the warm air, moves away to be replaced by cooler air. Finally, there's evaporation. When we sweat, the evaporation of water from skin carries heat away.

If needed, the body will widen blood vessels to pump more blood near the skin, a process called vasodilation, which releases more heat and thereby cools the body. (Cold conditions cause the opposite response, vasoconstriction, in which the body tries to preserve its heat by limiting that near-surface blood flow.) As the surrounding environment gets hotter, the body works harder to dump heat into the air around it and stick to its ideal temperature. This can lead to a condition known as heat exhaustion, in which the body sweats so much in its efforts to cool itself that dehydration and a loss of electrolytes results. The symptoms of heat exhaustion range from weakness and dizziness to headaches, nausea, and even vomiting. A person suffering from heat exhaustion needs to rest in a cool area and drink cool water to replace the fluid lost through sweat.[112]

The next stage after heat exhaustion is a dangerous condition called heatstroke. In heatstroke, the body loses its ability to control its temperature and cool off. The body's core temperature will rise to 105 or higher. A person affected by heatstroke can display confusion or slurred speech, have seizures, or lose consciousness. Heatstroke is a life-threatening condition that requires emergency medical treatment, which can include measures such as cooling the body with water or an ice bath.

But heat exhaustion and heatstroke tell only part of the story. They represent the direct impact of a person's exposure to high heat. The indirect effects can also be hazardous. When the body is forced to work hard to cool itself, it stresses the entire system, exacerbating medical conditions such as high blood pressure, heart disease, diabetes, and asthma. Wellenius told me how this increases the difficulty of gauging injuries and fatalities from a heat wave. "You may see a number of heatstrokes in emergency departments, and dehydration," he says. "But you also see many more people going to ED getting a diagnosis of renal problems or kidney failure or respiratory diseases. The larger impact across the population is a number of conditions that aren't directly identified as due to heat." The result: Many heat victims aren't necessarily recorded as such.

Some people are particularly vulnerable. Wellenius divides them into two categories. "One is people who are less resilient," he says. That includes not only people who have medical conditions but also people on medications that affect the body's fluid balance or blood flow; pregnant women, whose bodies are working overtime; kids, whose physiologic mechanisms aren't fully developed; and the elderly, whose bodies may have less reserve to draw on and who are more commonly on some of the medications that can be a factor.

Then there's a second category: People at higher risk from the heat because they're the most exposed to it. Think of people whose jobs require outdoor work, such as construction workers and landscapers, and delivery drivers, who spend much of the day in trucks that lack air-conditioning. Income disparities come into play here too, says Wellenius: "People who don't have to be outside for work but are waiting for the bus and there's no bus shelter. People who can't afford AC or can't afford to run it or can't afford to stay home all day." And of course, there are homeless individuals who spend most or all twenty-four hours of the day outdoors.

How hot is too hot for those at risk? That is yet another seemingly simple question about heat with a complicated answer. But understanding heat terminology in weather reports represents an important aspect of weather literacy. When it comes to heat's effect on the human body, there's more to consider than the figure on the thermometer. The number that we think of as "temperature"—the number you might hear on the TV or radio forecast or see on your phone—is the temperature of the air.[113] National Weather Service standards call for a temperature-reading device to be mounted about four to six feet above the ground in an open area but shaded from direct sunlight.[114]

For human comfort and health, another number comes into play: humidity. Remember that evaporation of sweat is one of your body's cooling mechanisms. Humid air is already packed with moisture, which reduces the potential for your sweat to evaporate. Everyone knows the old saw "It's not the heat, it's the humidity." Really, it's the heat *and* the humidity. In 1979, Robert Steadman, a scientist in Colorado State University's textile and clothing department, published a paper titled "The Assessment of Sultriness."[115] In his introduction, Steadman wrote: "Although comparisons of hot-arid with warm-humid climates are often the subject of a spirited exchange of opinions, little attention has been paid in the literature to providing an objective basis for such assessments. These comparisons require a single measure of the combined effects of high temperature and humidity."

Steadman wanted an indicator that would do for heat what the existing wind-chill scale did for cold—a way to express the *apparent* temperature that reflected what the conditions actually felt like. Given his work with textiles, he wanted to factor in not only the human body's experience of heat but specifically that of the clothed human body. The result is what we know today as the heat index. It's a chart with temperature along one axis and relative humidity along the other. You can find NWS's working version of this chart online.[116] Look up, for instance, 90 degrees and you will conclude that a day with the low-humidity value of 40 percent will feel like 91 degrees—but the same temperature with soggy 80 percent humidity will feel like 113 degrees. Note that *relative* humidity is used here, as it is in most everyday discussions of weather. Relative humidity describes how much moisture is in the air as a percentage of the maximum the air could hold at its current temperature. (Another measure is *absolute* humidity, which tells the actual amount of water in a given volume of air. But it's much less commonly encountered; usually when people say "humidity," they mean "relative humidity.")[117]

The heat index isn't the only approach to conveying the feel of heat. Canadian meteorologists use a measure called the humidex, which relies on the dew point instead of humidity. Dew point is the temperature to which the air would need to be cooled to in order to become saturated with water. Any cooler, and water in the air will condense out as dew. Humid air has a higher dew point, so both the heat index and the humidex are ultimately tied to both temperature and humidity.

Had enough? Here's another layer of complexity. I've used the word *temperature* to describe the reading on an outdoor thermometer or sensor. But in the heat index and humidex tables as well as the temperature forecasts you hear, the figure in question is actually the dry-bulb temperature. There is also a wet-bulb temperature, which is taken by a thermometer wrapped in water-soaked fabric, the idea being that, depending on the humidity, evaporative cooling from the water will affect the result. As with the heat index and the humidex, the goal is to better reflect the apparent temperature; all depend on both heat and humidity. There's more: a measurement called wet-bulb globe temperature (WBGT). This factors in the temperature reading from a sensor placed inside a black globe and exposed to the sun. Solar radiation gets absorbed by the black-coated globe. A sunny day will result in a high reading, providing a proxy for the heat absorbed by a human exposed to those hot rays, essentially a better way to tell the temperature for a person in direct sunlight. WBGT is often used as a scale to set health standards for activities outdoors during hot weather. Some high-school and college sports groups use WBGT in their recommendations to prevent heat exhaustion and heatstroke for athletes. The US Occupational Safety and Health Administration recommends the use of WBGT for outdoor workers.[118]

If you're still waiting for a simple answer to how hot is too hot, you might guess by now that one is not forthcoming. But it's worth digesting the National Weather Service overall guidelines, which call for caution when the heat index is 80 to 90 degrees Fahrenheit and extreme caution at 90 to 103 (heat exhaustion is deemed "possible" in that range with "prolonged exposure or physical activity").[119] Ratchet up to 103 to 124 degrees and you're in the danger bracket, and at 125 or higher, it's extreme danger, with "heatstroke highly likely." To put those numbers in perspective, a 90-degree day with 70 percent humidity amounts to a heat index of 105, already into the danger zone. A 98-degree day with 65 percent humidity has a heat index of 128, worthy of the extreme-danger caution.

There are several varieties of official NWS proclamations regarding heat. An excessive heat outlook can be issued "when the potential exists for an excessive heat event in the next 3–7 days." The next step up is a heat advisory, which NWS says can be "issued within 12 hours of the onset of extremely dangerous heat conditions." An excessive heat watch indicates "conditions are favorable" for a heat event in twenty-four to seventy-two hours. The most extreme notice is the excessive heat warning, which the NWS describes as something "issued within 12 hours of the onset of extremely dangerous heat conditions." The broad criteria for an excessive heat warning is "when the maximum heat index temperature is expected to be 105° or higher for at least 2 days and night time air temperatures will not drop below 75°." But here, too, it's complicated. Human physiology is a big variable. The bodies of those who live in hotter parts of the country are more acclimated, so the hazard level can vary based on location.

To understand this better, I talked with Paul Iñiguez, the science and operations officer in the Phoenix forecast office of the National Weather Service. As you might imagine, a meteorologist based in Arizona knows heat. The key to effective warnings is alerting people to the unusual, not the usual. Heat that's considered high for Seattle might be essentially the norm for Phoenix. To address this problem, Iñiguez worked with colleagues in the Western Region of NWS to develop an indicator called

HeatRisk.[120] It's a color-coded guide to heat dangers over the upcoming seven days, ranging from green ("Little to no risk") through yellow, orange, and red and topping out at magenta ("Extreme—this level of rare and/or long-duration extreme heat with little to no overnight relief affects anyone without effective cooling and/or adequate hydration"). "What HeatRisk does is take our forecast high and low temperatures and compares that to the climatology for that location for a given time of year," Iñiguez explains. It's still an experimental system, meant to provide context beyond the watches and warnings. "HeatRisk complements the heat warnings that go out. Because a warning is a binary thing, right? It's a warning, or there is no warning." Health is an important component. In developing HeatRisk, meteorologists consulted with the Centers for Disease Control to review health data that indicated what levels of heat in the region were likely to correlate with illness or death.

To Iñiguez, as with so many others I talked with, heat waves are no longer primarily a forecasting challenge. "These are big, robust systems, and they're really well forecast by global models now," he says. "Rarely is it a surprise anymore that we have any of these really big heat events." The problem isn't predicting the heat or even getting the word out. The challenge is making sure people understand just how dangerous this weather can be and ensuring they have some means to keep cool.

For an example, look no further than the deadly heat wave that devastated the Pacific Northwest in 2021.

It started with pressure. The signs were there in the computer-forecast models more than a week in advance. On Saturday, June 20, 2021, the National Weather Service's Spokane weather forecast office tweeted out an image of a weather map, specifically a 500-millibar (mb) chart. The 500 mb chart is one of meteorology's staple tools. It's a contour-style map that shows the height at which atmospheric air pressure will be 500 millibars. Standard air pressure at sea level is 1,013.25 millibars, so a 500 mb map is designed to display the altitude at which pressure is roughly half of sea level. This is important because areas of high and low pressure compose the major variables in atmospheric circulation . . . better known to you as weather.

Think of it this way: The air pressure you're experiencing at this moment is the weight of all the air above you in a column reaching all the way to space being tugged down by the Earth's gravity. The higher you go in the atmosphere, the less air you have above you; the weight of the column above decreases. It's why airliners require pressurized cabins. As students learn in high-school physics, there's a relationship between temperature and pressure. Heating a gas in a constant volume increases the pressure. Likewise, if you up pressure at a constant volume, temperature must rise. The 500 mb map is helpful because it highlights the areas of low and high pressure in the atmosphere and because the height of the 500-millibar level has a relationship to the temperature of the air below it. The hotter the temperature near the ground, the higher the 500 mb level will be. Meteorologists looking at the shape and relative locations of high- and low-pressure areas can draw conclusions about how quickly or slowly the weather in a given place will shift; generally speaking, high-pressure conditions mean it will take longer for the weather to change. Computer-forecasting models put out 500 mb maps that predict the location of those high- and low-pressure areas in the days ahead.

When the Spokane WFO tweeted out the computer-generated 500 mb map that Saturday in June 2021, it noted that it portended "record setting heat for next weekend." In another tweet, the forecasters added: "We know this forecast is still 7 days away, but we thought we'd give you a sneak peak at our latest forecast for next Sunday. The record for Spokane on 6/27 was set in 2015. It hit 102°F that day. Lewiston's record is 108°F also set in 2015."

The high heat was still a week away, but forecasters in the Pacific Northwest were already warning of record highs. In the days ahead the warnings ramped up—and so did the temperatures.

Two days after those tweets, on June 22, I spoke with Alex Lamers. He is the warning coordination meteorologist at the Weather Prediction Center (WPC) in College Park, Maryland. This particular outpost of the NOAA empire has a somewhat confusing name. After all, practically every part of the NWS organization is devoted in one way or another to weather prediction. In the case of the WPC, the focus is on national forecasts, precipitation forecasts, and guidance, generally out to seven days.[121] (One can compare that with the similarly named Storm Prediction Center, which concerns itself mainly with convective weather events—thunderstorms and tornadoes.) For instance, each weekday, the WPC issues a Day 3–7 Hazards Outlook with a color-coded map of the United States and overlays of various threats, from heavy rain or snow to high winds.

The Day 3–7 Hazards Outlook underscores the continuing role of the human forecaster when it comes to public safety. Numerical weather-prediction computer models churn out temperature predictions as a matter of course. But because no specific model is perfect, forecasters at the WPC examine the data from a variety of models, including ones operated by weather agencies outside the United States. They also look at ensemble forecasts. Accepting that there is no such thing as perfection when it comes to the input data that launches the model on its calculations, an ensemble forecast can run the simulation multiple times, slightly varying the initial conditions. By looking at the various ensemble members, a forecaster can see that some outcomes crop up again and again while others are outliers. Noting where those outcomes clump together is a helpful clue. For an imperfect analogy, think about cooking the same recipe again and again while slightly varying the initial ingredient amounts. You might find that most of the slight variations result in essentially the same dish, but some—for example, an extra pinch of cilantro or cayenne—produce a sharply different taste. Taken together, the group of different variations is the ensemble, and any individual forecast projection within the group is referred to as an ensemble member.

"One of the strengths of the WPC is expertise looking at ensemble models," Lamers tells me. "You're not relying on one or two forecasts. You're getting a full picture of the level of uncertainty that exists at that time range. There are a lot of factors that can evolve over the course of a week, so it's important to consider that full range of uncertainty." When it comes to alerting the world about heat hazards, Lamers explains, it's not simply a matter of looking at one model's forecast temperature for, say, six days into the future. "If you're relying on one specific model, it's going to lead you astray. But if there are, say, a hundred ensemble members and ninety of them on day six are showing you an extreme heat event, you can feel fairly confident."

When I spoke with Lamers, on June 22, the Day 3–7 Hazards Outlook map displayed a red line drawn around all of Washington State, nearly all of Oregon except for its southeast corner, and a chunk of northwest Idaho.

"Excessive Heat 6/25–6/29," stated the map legend. Lamers sounded confident about the high heat to come. "There's usually a pretty strong signal for these large-scale heat events. Right now in the Pacific Northwest, most of the ensemble members have a very anomalous ridge of high pressure aloft, which is a very strong signal for heat closer to ground-surface level," he said. "If we had lower confidence and higher uncertainty, we'd probably refrain from outlining that area on the hazards map."

That one WPC map represents nearly unimaginable amounts of computer data crunched into dozens of ensembles at forecast centers in both the United States and abroad, then interpreted by experienced meteorologists. The hazards map is meant to boil down all that and highlight key takeaways for possible dangers. The WPC describes the audience for the Day 3–7 Hazards Outlook map as including "emergency managers, weather forecasters, planners and managers in the public and private sectors, as well as the general public."

Later that Tuesday, June 22, the NWS Seattle forecast office sounded an alarm. "We've got a potentially record-breaking heat wave on the way this weekend into early next week. An EXCESSIVE HEAT WATCH is now in effect beginning Friday on into early next week," the Seattle WFO said on Twitter. In *The Seattle Times* the next day, a headline blared: "Brace for 100-Degree Days as Heat Cranks Up." The paper noted that the Emerald Downs horse track in Auburn had already canceled races for Sunday—four days away—due to the forecast high of 102 degrees.

In the latter half of that week, anyone living in the Pacific Northwest who paid any attention at all to traditional news outlets or social media would have been hard-pressed to escape word of the broiling weather ahead. The tone of impending crisis was unmistakable. In Portland, county health officer Jennifer Vines characterized the forecast temperatures as "life-threatening" and recommended everyone make sure they had a plan for where they could go to stay cool, according to *The Oregonian*. Officials across the region publicized the availability of cooling centers and underscored the vulnerability of homeless individuals. Restaurants, already battered by the COVID-19 pandemic, faced tough choices over whether to close or stay open and potentially expose kitchen workers to perilous heat levels. In Seattle, transportation officials worried about drawbridges becoming jammed if their metal structures expanded too much from the hot weather; they announced plans to shut down some spans so that they could be sprayed with cooling water, *The Seattle Times* reported.

The type of weather event that was unfolding is often referred to as a heat dome. The people of the Pacific Northwest were about to hear a lot about heat domes. Again, it all comes down to pressure. The jet stream, a high-altitude river of air moving from west to east, buckled in a certain way. It's best visualized as a length of string that someone poked from the bottom to form an upside-down U. If you could look down on Earth from orbit and see the jet stream, you would notice the river of air curving sharply upward to the north and then arcing back down to the south before resuming its flow toward the east. Within that open loop, or upside-down U, warm air became trapped. Scientists refer to this as a blocking event. In this case, it was called an omega block because the loop was shaped roughly like the Greek letter omega (Ω), with an area of high pressure nestled inside the loop of the omega. This blocking configuration can keep the high-pressure zone fairly stationary for several days. The warm air trapped in the high-pressure area tries to rise, but on its way up, it hits a lid from all that pressure. The warm air is now effectively trapped under a dome. Going back again to basic physics, increasing the pressure of a gas in a given space heats it up, and that's exactly what happens to the air trapped in the dome. The hot air lingers in place until the pattern eventually breaks up and the jet stream unkinks itself.

By Friday, the temperatures were climbing into the 90s and the NWS forecast office in Seattle was proclaiming the days ahead would be an "unprecedented event." On Saturday, thermometers climbed again. In Spokane, the high was 98. In Seattle, it spiked to 102. In Portland, the recorded high was 108 degrees. Hospitals were coping with emergency department visits from heat-related maladies. Then Sunday saw the heat ratchet up further, with highs of 102 in Spokane, 104 at Seattle-Tacoma International Airport, and 112 degrees in Portland. The TriMet public transit authority shut down its MAX trains, a streetcar-style light-rail network crisscrossing the Portland metro area. "The MAX system is designed to operate in conditions up to 110 degrees," the transit agency said. "Forecasts show it will likely only get hotter tomorrow without sufficient time to cool down."

Indeed, things were about to get worse. On Monday, June 28, the

National Weather Service recorded a high of 108 degrees at Sea-Tac airport, the hottest day ever there. Portland hit an all-time high of 116 degrees, shattering the previous record of 112 degrees that had been set just the day before. Spokane reached 105, and the electric utility company there, Avista, cautioned it would implement rolling blackouts to cope with stress on its system from the heat and the unprecedented demand for power to run air conditioners.

Tuesday, June 29, the heat finally begin to break, with Portland settling down to a high of 93 and Seattle dropping all the way into the 80s—though further inland, it took a few more days before temperatures fell out of triple digits. Spokane posted its all-time record high of 109 degrees that Tuesday and suffered through 104-degree weather on Wednesday before a respite on Thursday.

As the immediate emergency subsided, the extent of the damage revealed itself tragedy by tragedy. One day, King County officials reported two deaths from heat exposure; a day later the number of fatalities had grown to thirteen, and officials warned that count would surely increase. Details were often heartbreaking. *The Seattle Times* reported on the loss of Debra Moore, a sixty-eight-year-old woman who had succumbed that Monday after a sidewalk fall in Enumclaw, a small city east of Tacoma. Police there told the newspaper that the woman, who had significant pre-existing medical conditions, might normally have been found quickly, but due to the heat, people in the neighborhood had window blinds closed to shut out the sun. "She wasn't discovered there for quite a while, so she was laying there in the heat," an Enumclaw police spokesperson explained. In the days that followed, the death toll kept climbing. By July 8, at least seventy-eight fatalities in Washington State had been attributed to the extreme heat.

For the medical professionals and emergency responders who dealt with the surge of heat injuries, the heat dome proved unforgettable. But for Dr. Jeremy Hess, it was something more: a painful example of how unprepared society is to meet the challenges of extreme weather events. Hess, a plainspoken man in his forties, has an unusual viewpoint. He's a board-certified emergency medicine doctor and a professor at the University of Washington's medical school. He's also done extensive research on public health and the effects of climate change—particularly heat.

During the worst days of the heat dome, Hess was on duty in the emergency department of Harborview Medical Center, a hulking complex just on the other side of Interstate 5 in downtown Seattle. Harborview is the state of Washington's only level I trauma center, meaning it is staffed and equipped for every conceivable patient emergency. (It also happens to be an inspiration for the fictional hospital in the TV show *Grey's Anatomy*.)

"We started getting some warning of this event about five days in advance," Hess recalls. "From my point of view, five days was perfectly adequate to get the word out to people about this event. And I don't think the failures that we witnessed were failures about forecasting or the meteorological community getting the information out to people, I want to be really clear about that." (For context, five days out was when the National Weather Service in Seattle cautioned about the possibility of 100-plus highs on the way; the next day it warned of a "potentially record-breaking" heat wave.)

Despite the warnings, Dr. Hess was not sanguine about how the people of Seattle would cope. "When I heard those warnings, I thought, one, *People are going to die here*. And I thought, two, *I wish the risk communication about this event was more like risk communication about hurricanes and other severe weather events*." What he meant, Hess explains, was that the language people typically hear from a variety of sources—broadcast news, public officials, and such—prior to something like a hurricane underscores that the risk is catastrophic. Communicating the danger from the slow-motion catastrophe of extreme heat is more challenging.

"I was working overnight in the ED [emergency department] and the second night, there were some dramatic presentations related to heat exposure, but the ED was not overwhelmed by it. It was just your standard busy overnight shift in a hospital ED," Hess says. "The third night was different." Day three of the heat event was when Seattle recorded its all-time high of 108 degrees at Sea-Tac. "We got a lot of calls from EMS . . . about patients who were very sick, and who we were expecting. Several of those patients never got to us because they died before they got to the ED."

For patients who did make it to the emergency department, Harborview doctors and nurses filled body bags—those long plastic pouches normally used for deceased individuals—with ice. Heatstroke patients could be zipped into these makeshift coolers to help lower their dangerously high body temperatures. Hospital providers and emergency responders alike were witnessing conditions they'd never seen before. Amy Lee, a Harborview charge nurse, recounted that the pavement itself became dangerously hot. Speaking in August 2021 in a panel discussion for the Global Consortium on Climate and Health Education based at Columbia University, Lee said: "I remember specifically speaking with one paramedic, and I asked him how bad it was out there. He showed me where he had little burns on his knees from being on the ground intubating [a patient]."[122]

Hess stresses that a contributing factor to the heat's deadliness was the region's lack of experience with temperatures so high. Only 44 percent of housing units in metropolitan Seattle had some form of air-conditioning (window units or central air) as of 2019, according to the Census Bureau's American Housing Survey. By contrast, over 80 percent of Los Angeles housing units had air-conditioning in the same survey.[123] Hess himself did not have air-conditioning at home; it's usually not needed in Seattle's temperate climate. When he worked overnight shifts, he would often sleep in his home's basement in the daytime for the quiet and dark during daylight hours—and during the heat event, it was at least a little cooler. Still, he recalls that when he climbed up the stairs after sleeping, he could feel the temperature rising halfway up. "You come just to the ground floor, and it was like a hundred degrees in our kitchen."

For Hess, as tragic as the toll from the 2021 heat dome turned out to be, the prospect of more—and worse—in the future is equally sobering. Speculation in the midst of the heat dome was that the extreme nature of the event must somehow reflect the impact of climate change. "I've seen a lot of tough things in my work as an emergency medicine doc, and this definitely stuck out," Hess says. "And it was discouraging, given that I also work in climate and health." He adds there was no question in his mind that it wouldn't have happened without climate change. "It was the first event that I've been in that I could totally, unequivocally, say that," he tells me.

In an unprecedented project, a group of climate scientists moved quickly to demonstrate the speculation was correct. Geert Jan van Oldenborgh, at the Royal Netherlands Meteorological Institute, and Friederike Otto, then at the University of Oxford, were cofounders of a project called World Weather Attribution and pioneers of attribution science—the work of scientifically attributing a given weather event to climate change. With the heat still baking the Pacific Northwest, van Oldenborgh and Otto led a group of scientists working across the world to analyze the heat dome. Only days after temperatures ebbed—on July 7—the team published their analysis, declaring the deadly heat was "virtually impossible without human-caused climate change."[124] They added, in a sobering warning: "Looking into the future, in a world with 2°C [2.7 degrees Fahrenheit] of global warming (0.8°C [1.44 Fahrenheit] warmer than today which at current emission levels would be reached as early as the 2040s), this event would have been another degree hotter. An event like this—currently estimated to occur only once every 1000 years, would occur roughly every 5 to 10 years in that future world with 2°C of global warming." The speed with which they produced these findings was part of the point. By getting their conclusions out so quickly, van Oldenborgh and Otto and their colleagues were able to publicize them while the Pacific Northwest aftermath was still in the news, the better to galvanize public attention. World Weather Attribution made use of peer-reviewed methodology, powerful computers, and the ability to spread work across time zones to accomplish in days what usually takes months or years. As a result of their work, van Oldenborgh and Otto were on *Time* magazine's list of the world's most influential people for 2021.[125]

Hess says the 2021 experience suggested the dangers moving forward. "I expect in the future we will have an event that's considerably worse," he says. The nightmare for Hess and other experts in public health is an event that lasts longer and becomes a layered-threat situation. Electric utilities and grids already struggle during heat waves due to high demand and the punishing effect on equipment. In a layered-threat scenario, the first emergency—hot weather—leads to another: a power outage that takes out air-conditioning, eliminating the ability of individuals to cool their homes and severely straining institutions and cooling centers even if they're equipped with backup power.

As the weeks passed in early summer 2021, the dimensions of the disaster came into sharper focus. In its final report, the Washington State Department of Health concluded there had been one hundred heat-related deaths between June 26 and July 2.[126] In late July, Oregon officials published an "after-action review" to assess the state's response to the heat wave. It put the statewide death total at eighty-three.[127] British Columbia, Canada, suffered tremendously during the heat event—Victoria, the provincial capital on Vancouver Island, topped 103 on June 28—and a report nearly a year later identified 619 heat-related deaths.[128] "Most of the deceased were older adults with compromised health due to multiple chronic diseases and who lived alone," the report noted.

The true death toll may never be known. Extreme heat kills in too many ways to make a precise accounting possible. It strikes some directly through heatstroke. It claims others indirectly by stressing their bodies and worsening existing health issues, such as diabetes or kidney disease. Some deaths come almost immediately; others come later as the body loses its battle. Epidemiologists can use statistical methods to look for anomalous patterns of mortality. But even though we cannot account for every single victim at an individual level, there is no question at this point about the danger to human health from extreme heat waves.

Richmond's streets begin to come alive as our Chevy Trailblazer roams the city and the sun edges higher in the sky. After about an hour, the morning run wraps up. Volunteers will take to Richmond's streets two more times this day. While the morning run was meant to capture data at the coolest time of day, the afternoon journey will record temperatures at their most punishing. An evening run rounds out the effort.

This single day's work represents a sizable undertaking. Twelve different routes have been devised to help drivers cover the city, ensuring that data will capture conditions in different neighborhoods that vary in their use and environment, from tree-lined residential streets to industrial areas blanketed by concrete and asphalt. Multiply those twelve routes by three, for the morning, afternoon, and evening collection runs, and the campaign amasses a tremendous amount of block-by-block data. Just determining the date for the event is a project in itself: Researchers want a hot and sunny day to ensure the most useful data. Fortunately, National Weather Service predictions on that front are generally quite reliable days in advance.

A key architect of this mapping effort and, more broadly, urban heat mapping in the United States is a passionate scholar of cities, climate, and the environment named Vivek Shandas. He's a professor in the urban studies and planning school at Oregon's Portland State University, where he founded the Sustaining Urban Places Research Lab. In 2014, Shandas and his team set out to learn which areas of Portland were disproportionately hot and then compare that information with demographic maps. Their research, published in 2018, found "significant relationships between heat exposure and populations that are low-income, non-white, minimally-educated, or poor English speakers; all of these socio-demographic groups, as well as those living in affordable housing, experience higher temperatures than their wealthy, white, educated, English-speaking counterparts."[129]

The technique that Shandas and his colleagues developed for that 2014 study—mounting sensors on vehicles and sending them to traverse city streets—became the basis for similar campaigns across the United States. As interest in mapping urban heat islands grew, Shandas launched a firm, CAPA Strategies, to expand the work, advise communities about climate issues, and offer tools to respond. (CAPA is an acronym for "Climate, Adaptation, Planning, Analytics.") With support from NOAA and the National Integrated Heat Health Information System (NIHHIS, usually pronounced "*nye*-hiss"), a joint program of NOAA, the Centers for Disease Control, and other agencies, Shandas and CAPA have helped enable heat-mapping campaigns in more than fifty US cities, from San Francisco and Albuquerque to Baltimore and Boston.[130] The playbook involves more than just plotting route maps and providing sensors. It places community organizations squarely at the center of the effort. Local groups can apply to NIHHIS for the opportunity to host a campaign, and volunteers drive the routes.

When I talked with Shandas, he underscored the value of putting local leaders and organizations in key roles. His points echoed what I heard about Hunting Park's experience with its planter/bench program: Get more people involved, raise awareness about health dangers, and draw attention to weather and climate inequities. "How do we localize this big planetary climate issue?" Shandas says. "We are socializing these conversations with groups that may not have thought about why it has been one hundred and ten degrees for five days straight." The novelty of the mapping efforts combined with the involvement of local volunteers helps attract local media attention; newspapers, TV, and radio frequently cover the mapping campaigns as an event but wind up also covering the broader issues of heat islands, heat health dangers, and environmental justice.

In 2017, Richmond conducted its first heat-mapping campaign. The effort was staged by a partnership that would prove consequential for our understanding of heat's disparate impact on different neighborhoods. Shandas teamed up with Jeremy Hoffman, then a climate and earth scientist at the Science Museum of Virginia in Richmond. Hoffman recalls meeting Shandas for breakfast in Portland in the summer of 2016 after being introduced by a mutual friend. "I was so enamored of his work," Hoffman tells me. He wanted to try a similar project back home in Virginia but realized it would be complicated and costly for Shandas to bring his entire team out east. That's when they hit on the idea of local volunteers (or, as they began to refer to them, "citizen scientists") from the community who would drive the sensor-equipped cars around the routes. "Little did I know that was going to be the thing that makes these campaigns, beyond the technology, beyond the methodology, which is state of the art," Hoffman says. "But it was that community angle that was going to elevate it."

The Richmond results were fascinating. More than just adding to the evidence of how widely temperatures vary across a city due to the heat-island effect, they also pointed to how often the hot spots can be found in neighborhoods with minority or lower-income residents.

This information came from another kind of map: the geography of redlining. After the passage of New Deal legislation that put the federal government in the role of supporting housing programs and mortgage lending, the Home Owners' Loan Corporation (HOLC) was tasked with creating so-called residential security maps for cities across the country. The stated purpose was to help banks assess the risk of home loans in different areas. The maps used red to signify what were claimed to be the riskiest neighborhoods, often areas with the highest concentrations of Black residents and other minorities. Those maps became the foundation of—and gave their name to—the practice of redlining.[131] When banks engaged in redlining, they withheld or applied restrictive conditions to loans for those neighborhoods, something that is now understood to have perpetuated housing segregation, with effects that linger to the present day.

Hoffman and Shandas realized they could take the old HOLC maps and compare them with heat maps. To cast their net wide and ensure consistency across different cities, they didn't rely on ground measurements. Instead, they drew on observations from Landsat 8, a satellite launched in 2013 that can read infrared wavelengths that can then be used to determine surface temperatures. What they found: Significant correlation between higher heat and the HOLC redline maps. "Our results reveal that 94% of studied areas display consistent city-scale patterns of elevated land surface temperatures in formerly redlined areas relative to their non-redlined neighbors by as much as 7°C [12.6 degrees Fahrenheit]," Hoffman, Shandas, and colleague Nicholas Pendleton wrote in the paper that appeared in the peer-reviewed journal *Climate*.[132] "Nationally, land surface temperatures in redlined areas are approximately 2.6°C [4.7 degrees Fahrenheit] warmer than in non-redlined areas." What explains the difference? Factors include the relative amount of tree cover and parkland versus roads and rooftops. But ultimately, the point was to demonstrate the connection between where the heat is worst now and the historic roots of housing discrimination. The findings attracted national coverage, including stories by NPR and *The New York Times*.

I met Hoffman in person for the first time on the eve of Richmond's 2021 mapping campaign. The Science Museum of Virginia building is a neoclassical beauty with an imposing colonnade entrance and a soaring five-story rotunda with a Foucault pendulum that swings to and fro, knocking over pegs in a proof of the Earth's rotation. Hoffman walked me out to his car and took me on a driving-tour introduction of Richmond as a preview of the mapping routes I would travel with volunteers the next day.

It was fascinating to see Richmond through Hoffman's eyes, acutely tuned to the role of heat in the city. As Hoffman's Subaru made its way down Commerce Road, a major thoroughfare in South Richmond, Hoffman pointed out a man near a bus stop. Rather than standing by the curb, the man was on the opposite edge of the sidewalk, pressed up against an embankment wall supporting an adjacent parking lot on a hill. "He's trying to get some shade in the afternoon," Hoffman explained. Once he pointed it out, I could see how the angle of the sun meant that the retaining wall offered a narrow respite from the direct sunlight. It turns out there's a tie-in to the heat-island mapping here too. "It has precipitated a conversation around access to shade," Hoffman told me. "Many people that are living in hotter areas also tend to rely on public transportation as their primary form of how to get around the city." Thanks to heat-mapping data, it's now possible to place shaded bus stops where they're most needed.

So how do we solve the problem of dangerous heat? The gravity of the threat just keeps growing. The 2021 heat event in the Pacific Northwest was followed by the summer of 2022's extreme heat in Europe, with London reaching 104 degrees and one part of Portugal reaching 117 degrees and eventual estimates of deaths in the tens of thousands.[133]

Of all the varieties of dangerous weather, heat is perhaps the least likely to be helped by better forecasts, simply because the forecasts are already so good. NOAA's Weather Prediction Center has a remarkable record of accuracy and improvement. For 2020, its day 3 predictions for high temperatures were off by an average of about 3 degrees. Back in 2000, that error was a bit shy of 4.5 degrees. As for longer-term forecasts, the day 7 error for 2020 was just over 5 degrees, more accurate than a day 3 forecast from thirty years ago.[134] There's every reason to expect further gains in accuracy as computer weather-prediction models continue to improve. And in the case of extended events like the 2021 heat dome, forecasts are able to see the high temperatures coming because the atmospheric blocking—that buckling of the jet stream that parks a high-pressure area in place for days—is a major atmospheric feature that stands out.

The trouble with heat isn't knowing when it's on the way. Meteorologists have mostly got that licked. The problem is doing something about it. Forecasts often fail to produce better outcomes. In 2021, the people of the Pacific Northwest had days of urgent notice about the coming heat; local National Weather Service offices used words like *unprecedented* to describe the high temperatures on the way and publicized the heatstroke symptoms that would warrant calling 911 ("throbbing headache, confusion"). As far as meteorology is concerned, the alarms are being sounded.

There are no simple fixes. At an individual level, people can prepare by installing air conditioners—assuming they can afford to purchase them and pay for the power they consume. (A cheaper alternative are so-called swamp coolers, evaporative coolers that blow room air through water-soaked pads, but they don't provide much comfort in high humidity. Still, because they actually cool the air, they're better than fans, which only move the hot air around.) Those who lack air-conditioning can shelter at a cooling center, but many are reluctant to do so or may not appreciate that the heat health threat warrants it. There are some government and nongovernment programs that provide air conditioners, such as the state of New York's Home Energy Assistance Program. In Philadelphia, utilities are barred from disconnecting customers (because of, for example, failure to pay bills) when the city has declared a heat health emergency. But overall, efforts to ensure that people have access to cooling have lagged behind those that protect access to heat in winter.

More broadly, we need to get people to take the threat of heat seriously before it's too late. That's one reason why programs like the shade planters in Philadelphia and the heat-mapping campaigns conducted in Richmond and other cities matter beyond their specific results. Efforts like these help raise the volume of the conversation about heat, emphasizing the health dangers and the ways in which some neighborhoods and populations are disproportionately at risk.

It's important to keep those specific groups in mind. Children and the elderly are obvious targets for increased awareness, given that their physiology puts them in extra peril. Similarly, people with chronic health issues, such as diabetes and heart disease, must take care because of the stress that severe heat can place on their bodies. Homeless people present a particularly difficult challenge, making outreach efforts and availability of cooled shelters critical. And finally, outdoor workers—who are not only exposed to the heat but also often engaged in strenuous activities—need better protection. In 2022 the Biden administration launched an effort to scrutinize conditions for these workers, with the Occupational Safety and Health Administration (OSHA) stepping up inspections and enforcement.[135] But worker advocates say what's really needed are clear federal rules that require employers to take specific steps to protect employees. OSHA is in the process of developing these rules, but some fear it will drag on due to employer opposition.

Longer term, obviously, looms the opportunity to address climate change by reducing carbon emissions—no small task. But there's growing acceptance—part of the broader reckoning with the changing climate—that the need to respond to heat will grow whether you're in Portland or Phoenix. In the meantime, cities can work to adapt to higher heat by encouraging buildings with roofs that are "green" (covered with plants) or "cool" (composed of materials that reflect sunlight rather than absorbing it). Increasing tree cover and parkland can also help, particularly for the hottest areas. In the wake of the original 2017 heat-mapping project in Richmond, Mayor Levar Stoney announced a project to convert unused parcels of land owned by the city into neighborhood parks, and the city council approved an ordinance in 2020 to create five new parks in Richmond's Southside neighborhoods, home to some of the most prominent heat islands. Richmond also has a broader effort called RVAgreen 2050, which the city describes on its website as an "equity-centered climate action and resilience planning initiative to reduce greenhouse gas emissions 45% by 2030, achieve net zero greenhouse gas emissions by 2050 and help the community adapt to Richmond's climate impacts of extreme heat, precipitation, and flooding." Cities can also follow the lead of Los Angeles and Miami–Dade County, which have created chief heat officer positions to prioritize climate and heat-resiliency efforts.

An event like the 2021 heat dome can serve as an effective, if grim and belated, driver of change. About a year after it, King County, which includes Seattle, announced a multiagency effort to develop the county's first-ever extreme heat mitigation strategy. The announcement was highly climate-conscious; it said the strategy "will identify actions needed to enhance the region's immediate response to extreme heat while adapting the built environment so that people and property are better prepared for more prolonged, hazardous heat waves that climate scientists predict" unless there is a dramatic increase in greenhouse gas emissions.[136] King County officials also outlined steps already underway, including better health-centric communications to the public on heat and disseminating emergency warnings in nine different languages, up from two. Longer term, county officials foresee increasing tree cover and supporting energy-efficient housing. Transit officials will work to deploy bus-stop shelters using urban-heat-mapping data from a NIHHIS/CAPA campaign.

Another idea is assigning names to heat waves, copying the well-known process of naming hurricanes. A nonprofit focused on climate issues, the Adrienne Arsht-Rockefeller Foundation Resilience Center (since renamed the Climate Resilience Center), has advocated this approach. In July 2022, a group in the city of Seville in Spain announced that it was designating a dangerous heat event then underway as Heat Wave Zoe. The effort involved local officials, university researchers, and the Arsht-Rockefeller group. The hope is that using nomenclature associated with hurricanes might better emphasize the threat. And in an age of social media, an event name can simplify and fuel online discussions. But the approach isn't without controversy; some fear it can create confusion with the hurricane system. In the United States, the Weather Channel assigns names to major winter storms, but NOAA and the Weather Service have not endorsed the idea.

Though it's important to test these types of innovations to determine their effectiveness, it's encouraging nonetheless to see more out-of-the-box thinking applied to heat forecasts. Some cities are working to move past bare-bones cooling centers to more elaborate facilities (*resilience hub* is one of the terms favored). In Arizona, the city of Tempe opened a building that serves as a combination community center, refuge, and all-around service facility, complete with ice supplies for residents when needed. As we face a future where the deadly and invisible threat of extreme heat keeps growing, we will need all of the above and more. It's only going to get worse. Cities, counties, and states all need to respond with infrastructure and urban planning that can cool our neighborhoods and bring relief to all who need it during emergencies. Organizations from schools to employers will need their own plans. Big community efforts like these are critical. Society needs to demonstrate that it can mobilize to combat heat the same way it reacts to hurricanes and tornadoes.

But it also strikes me that there's a potential for new types of individual warnings. We live in an era of both accurate temperature forecasts and nearly pervasive personal technology, at least for those who can afford smartphones. Our phones know where we are at all times. Fitness devices and smartwatches can monitor health indicators such as heart rates and pulse oximetry. Some wearables can already measure body temperature. It's not hard to imagine our personal technology providing us with not only the weather forecast but also detailed information on when we're entering a heat island where the actual temperature surges from the "official" reading. A smart app could even advise people on a cooler route to where they're going when they're out walking or alert them concerning rises in body temperature. If technology can give us algorithms that cater to our personal behavior when it comes to music or video clips, surely it should be able to counsel us on how to avoid the hazards of our environment.

A little after four p.m. during the day-long heat-mapping project in Richmond, the volunteers who hosted me for the afternoon data run drop me off downtown, our route complete. We've recorded the most scorching moments of the day. And once again, answering *How hot is it?* proves complicated. Our assigned route took us through Richmond's Gilpin Court neighborhood, on the city's north side. The layout of this neighborhood is a map of inequity. It takes its name from the New Deal–era Gilpin Court public housing project. It was originally part of Jackson Ward, a neighborhood with a rich legacy of Black residents and a stretch of Second Street known as Black Wall Street. But highway construction in the 1950s cleaved the neighborhood in two, with Interstates 95 and 64 cutting Gilpin Court off from Jackson Ward. Today, "Reconnect Jackson Ward" efforts are finally underway to redevelop the Gilpin housing and integrate the two neighborhoods.

After all the data is downloaded from the probes and analyzed, Gilpin turns out to be one of the hotter neighborhoods, with an average afternoon temperature of 91.48 degrees. In contrast, leafier (and wealthier) Stratford Hills, to the city's west, averages 89.49. But most striking is the variation across the entire city. The hottest point was 95.2, while the coolest was 86.6—more than eight degrees of difference.[137] Time and again, heat-mapping campaigns have found these kinds of significant variations, frequently with the hotter areas corresponding to lower-income neighborhoods. No one is immune to the effects of extreme heat. But when the forecast calls for sweltering temperatures, these are the communities that suffer most.

6 Hurricanes

A PLANET-WIDE VIEW TO TRACK DEADLY STORMS

Follow the River Thames upstream, west from its serpentine path through London and past Slough, and you'll arrive at the town of Reading. About a ninety-minute drive from central London (or, more enjoyably, a half-hour train journey from Paddington Station on the Great Western Railway), Reading boasts a busy, walkable shopping district where independent stores and traditional English pubs mix with American fast-food chains like Five Guys. Like so many towns in England, its history stands in plain sight. Walk a few minutes from Town Hall Square and you'll encounter the remains of Reading Abbey. The monastery dates back to the year 1121, when King Henry I, a son of William the Conqueror, founded it with royal grants of land. In 1453, Parliament sat in Reading. A plaque commemorates the attendance of a young Jane Austen at a school in the Abbey Gateway in 1785; she drew on the experience to portray the girls' boarding school in *Emma*.

In a building just a few miles away from this millennium-deep well of history, behind a secure door, are exemplars of twenty-first-century technology: two Cray XC40 supercomputer clusters, each capable of more than four *quadrillion* calculations a second. One bears the name Anemos, from the Greek word for "wind"; its counterpart is Ventus, which means "wind" in Latin. Their appearance doesn't betray their immense power. They're just rows of big metal cabinets; they look like a bunch of refrigerators placed side by side.

Unusually, though, these two supercomputers have been decorated with artwork across their cabinet fronts. The design incorporates some obvious emblems of meteorology, such as satellites and a balloon carrying instruments. It also includes some complicated-looking mathematical equations swooping over multiple panels. But the most significant element is a stylized map. Colored squiggles partly obscure the geography of a landmass, requiring a beat on the part of the viewer to recognize the territory. But after a moment, the map comes into focus—the outline of Florida, the Gulf of Mexico, and on up the Eastern Seaboard tracing the unmistakable contours of North America. Amid a tangle of lines, dots mark out a path beginning in the ocean north of Panama, slicing across Cuba, and curving north up the Atlantic Ocean. The dots trace a course that remains east of the US coastline before making a left turn and crossing over land in the vicinity of New Jersey.

The diagram tells the story of one of the most famous forecasts in modern meteorology, the 2012 track of Hurricane Sandy—and the building I'm in houses the organization that produced that notably accurate prediction. This is the European Centre for Medium-Range Weather Forecasts (ECMWF). I've traveled here to the ECMWF's Shinfield Park headquarters in Reading because the organization represents the gold standard for global-forecast models, the indispensable tools for predicting hurricanes days in advance. Here at the ECMWF, scientists oversee massively complex numerical weather-prediction (NWP) models and work to constantly improve them. Their success in doing just that translates to better weather prediction for people everywhere.

That's because global models serve as linchpins for practically every corner of modern weather forecasting. Day in and day out, they crunch enormous amounts of data to simulate how the Earth's atmosphere will behave in the near future. These pictures help meteorologists warn of heat waves on the way. They tell of low-pressure systems likely to bring rain or snow. They tip forecasters off to whether conditions will be cloudy or windy.

But the global models really get their moment in the limelight during hurricane season. They can project how big a hurricane will get and what its trajectory will be—critical intelligence for getting people out of the way of these tropical storms, which routinely result in destruction measured in billions of dollars and terrible loss of life. To take one recent example, Hurricane Ian in September 2022 cost an estimated $118.5 billion in damage, according to NOAA, and 156 people died.[138]

Worse, the threat appears to be growing in some ways as a result of climate change. Climate scientists expect future hurricanes to bring higher rainfall levels, increasing the danger from floods.[139] There's a simple relationship at work: warm air can hold more moisture than cold air, so a warming climate can be expected to increase hurricane rainfall rates—by something like 10 percent to 15 percent, according to a NOAA review of relevant research.[140] Some studies also project an increase in hurricane intensity, with a potential for fewer storms overall due to changes in atmospheric circulation that affect storm formation but with more storms reaching category 4 or 5 levels.[141] Add expected rises in sea levels—which play a key part in coastal flooding during tropical storms—and the outlook becomes even more alarming.

The story of hurricane forecasting is one of remarkable progress, largely thanks to the evolution of NWP models—much of which remains unappreciated by the public. Year by year, forecasts of hurricane tracks and strength keep getting better and more accurate. But it's also a story that must be reconciled with tragedies and inequities. When Hurricane Katrina smashed into Louisiana in 2005, the effect was catastrophic, particularly in some of the poorest neighborhoods of New Orleans. Nearly 1,400 people died.[142] The number of houses destroyed created a homelessness crisis that persisted for years, all from a single storm.

Just as with forecasts for tornadoes, heat waves, or any of nature's hazards, hurricane predictions require a spectrum of efforts to have good outcomes. It begins with long-term planning by officials and advance preparation by individuals and families, all conducted well before the danger from a specific hurricane. Then, when a storm materializes, people need guidance about what to do. Accurate forecasts, ones that officials and the public will put their faith in, can provide precious hours or even days to prepare or evacuate.

The work to provide those forecasts spans a broad range of the weather-prediction world. It starts with observations, the measurements of the atmosphere needed for a snapshot of weather at the moment. Next, those observations are fed into the computer models, where they serve as a starting point for calculations. Once the models have done the math and generated their expectations of the weather ahead, the computer-generated output must be translated into forecasts and warnings by expert human meteorologists who know how to interpret all this data. Finally, this information must get shared effectively with the public—most often by broadcast meteorologists, including some who are working on innovative ways to make sure that people in a hurricane's path understand the dangers and the need to take action.

To explore hurricane forecasting and, more broadly, the value of global computer models, I'll look more closely at each link in this chain. It involves all three elements of the weather enterprise: academic researchers improving our understanding of tropical storms, government-backed agencies running the computer models, and companies from the aerospace industry to the media. It's a complicated mix of public and private entities. That increases the urgency of making sure that science and the public interest drive the work, a requirement that has become more challenging due to the politicization of climate change, which has begun to encroach on weather forecasting as well.

Every component of this network matters. But numerical weather prediction is the backbone, which is why any discussion of hurricane predictions must center on the global computer models. The work to produce the best model may be one of the most intense competitions in meteorology. In the United States, NOAA has labored to improve its primary global model, driven in part by the dominance of the ECMWF's model. After years of watching the Europeans outdo their American counterparts, critics have suggested that NOAA needs to streamline its operations and work more quickly to incorporate improvements. Public safety benefits from these rivalries. When human forecasters, whether in the United States or abroad, weigh the information available to them, they consider the output from all the major models. But thanks to its consistently high performance, the ECMWF's model has essentially become the benchmark worldwide.

What makes the ECMWF so good? It's distinctive in several ways. One is evident when you leave the supercomputers behind and step into a wood-paneled room on the second floor of its Reading headquarters. Dominating the space is a large oval conference table with a microphone at each seat and a gavel at the table's head. This is the council chamber. In contrast with most weather organizations, from the National Weather Service in the United States to the Japan Meteorological Agency, the ECMWF serves not a single government but a consortium of twenty-three member nations and twelve cooperating states (as of early 2024). The center of the table holds a collection of small flags representing those members, which include the United Kingdom and most of the European Union countries. Among the cooperating states are Israel, Lithuania, and Morocco. A glassed-in mezzanine along one wall adds to the United Nations vibe; it accommodates translators when needed to facilitate a council meeting.

Another distinguishing feature stems from the consortium approach. The ECMWF isn't public-facing. It doesn't issue forecasts for a general audience. Instead, it focuses tightly on its mission of operating global NWP models and conducting research to improve them. Its "customers" are those member states, most of which have their own national weather agencies. In the United Kingdom, it's the Met Office; in Germany, it's the Deutscher Wetterdienst, or DWD. The meteorologists in each country use the ECMWF model's predictions to produce their own forecasts and then disseminate them to the public.

Yet in spite of that behind-the-scenes mission, the ECMWF has gained a considerable public profile, in no small part due to hurricane forecasts like the one that correctly foretold Sandy's landfall in New Jersey. If you watch TV coverage of a hurricane in the United States, you're likely to hear mention of what broadcasters have taken to calling the European model or simply the Euro, shorthand for the projections from the ECMWF. This visibility seems even more remarkable when you consider that the ECMWF's own member countries rarely need to worry about hurricanes; tropical cyclones in the Atlantic predominantly affect the Caribbean and the East and Gulf Coasts of the United States.

So with each new hurricane season, as the first tropical cyclones begin to spin up in the Atlantic, all eyes turn toward Reading. But to appreciate what makes these forecasts so valuable, you need to understand how a tiny disturbance in the atmosphere can blow up into the monster storms we know as hurricanes.

A hurricane begins life as a wave. Not a wave in the water, but one in the air. Meteorologists use the term *wave* in this context to describe a disturbance in the pattern of winds. When it comes to hurricanes, the winds that matter come from Africa. They blow from east to west, from the continent's landmass toward the Atlantic Ocean. If you picture wind as a river of air, you can imagine an individual wind current as a line superimposed on a map of Africa's west coast and the eastern Atlantic. Meteorologists use just that kind of visualization in what's called a streamline chart. A typical streamline chart focused off the coast of Africa will show a group of parallel lines, each representing a wind flow, with arrows indicating the east-to-west movement. Generally the lines remain the same distance apart as you follow them west. But sometimes a disruption will appear, created by a trough of low-pressure air perpendicular to the wind flow, making peaks in the otherwise smooth curves. Over time, that disruption will move across the streamlines from east to west, bringing rainy or stormy weather to the east of the trough.

That's a tropical wave. Hurricane forecasters keep close tabs on them. They are the small seeds that can eventually grow into an epic storm. The basic ingredients for a hurricane are warm ocean water, humid air, and the converging winds on the eastern side of a tropical wave. Another ingredient is the Coriolis effect, named for the French scientist Gaspard-Gustave de Coriolis, which refers to the way that air masses rotate as they travel above the Earth's surface: counterclockwise in the Northern Hemisphere and clockwise in the Southern Hemisphere. Because the Earth revolves on its axis, a point near the equator moves at a faster speed than a point near one of the poles. In the Northern Hemisphere, that differential in speed proceeding northward from the equator causes air masses heading toward the North Pole to deflect toward the east, or right, while air moving toward the equator deflects to the west. The result is that air traversing around the low-pressure center of the hurricane will rotate counterclockwise.

It's worth a brief detour here to explain some hurricane-related terms. A *tropical wave* marks the very beginning of a hurricane, and the next step is a *tropical depression*. That's when a rotating storm—a cyclone—forms with maximum sustained winds of thirty-eight miles per hour or below, according to NOAA's definition. The next stage is a *tropical storm*, with winds of thirty-nine to seventy-three miles per hour. When maximum sustained winds exceed seventy-three miles per hour, that's a *hurricane*. A tropical wave won't necessarily grow into a tropical depression, and any given tropical depression may peter out and not intensify into a tropical storm. All of these—tropical depressions, tropical storms, and hurricanes—are considered tropical cyclones. And just to make things more complicated, the powerful cyclones that are deemed hurricanes when they're in the North Atlantic or the eastern portion of the Pacific are referred to as typhoons in the northwest Pacific.

Why do some tropical waves grow up to be hurricanes while others don't? Scientists know those basic conditions necessary for a tropical cyclone to form—warm ocean water, moist air, the Coriolis force, and a few other atmospheric conditions—but they still strive to understand exactly why those ingredients sometimes result in a storm and sometimes don't. That's why the computer models, which simulate the future state of the atmosphere based on current conditions, have become such a valuable tool. What we do know for sure is what happens when a tropical cyclone *does* form. In their introductory textbook *Meteorology Today*, C. Donald Ahrens and Robert Henson explain: "A cluster of thunderstorms must become organized around a central area of surface low pressure."[143] When those storms start rotating around the low-pressure center, the nascent cyclone draws its power from the warmth of those tropical ocean waters. Air flowing toward the low-pressure zone—which becomes the eye of the storm—picks up heat from the ocean surface. As the warm air flows inward, it rises, bringing moisture into the clouds that surround the eye wall. As the cyclone grows, those surface winds blowing toward the center pick up speed, absorbing more heat from the water and intensifying the storms around the eye.

The result is the characteristic structure of a hurricane, familiar to most from satellite imagery: a rotating storm hundreds of miles wide with a relatively calm center (the eye) and alternating bands of rain and clearer skies twisting out from the eye. That warm ocean water serves as the fuel for the engine of the cyclone, which is why the storms begin to peter out once they move over land.

It's difficult to overstate the destructive potential for a major hurricane. In one often-cited comparison, NOAA says that, from start to finish, a single hurricane can release heat energy "equivalent to a 10-megaton nuclear bomb exploding every 20 minutes." Wind speeds get much of the attention in hurricane coverage, and they are truly fearsome. Meteorologists classify hurricanes according to wind speed on the Saffir-Simpson Hurricane Wind Scale, which dates back to the early 1970s. A category 3 hurricane blows at 111 to 129 miles per hour; think of the wind power of a good-size tornado. But whereas a tornado usually lasts a few minutes and carves a relatively narrow path of destruction, hurricanes can span hundreds of miles. The Saffir-Simpson scale goes up to category 5, which describes storms with sustained winds of 157 miles per hour or higher. In 2015, Hurricane Patricia, which reached category 5, unleashed winds over 200 miles per hour while moving over the warm Gulf of Mexico waters. For context, a Boeing 737 jet can achieve takeoff when air moves past its wings at about 170 miles per hour.[144] Winds at these speeds can rip the roof away from a home or destroy it altogether.

As much as wind-speed numbers dominate news coverage, talk to any hurricane expert and they'll tell you: It's the water, not the wind, that kills. That isn't to minimize the threat from those tornado-strength gales, but any given hurricane can easily dump ten or more inches of rain. If that much precipitation fell as snow, it would pile up ten feet or higher. Big storms get much wetter. Hurricane Harvey in 2017 deluged some locations in Texas with over sixty inches of rain. The storm effectively stalled when it reached the Lone Star State, hovering over the region for several days, absorbing warm Gulf of Mexico waters and releasing them over land as rain. In Harris County, which includes Houston, flood-control district officials later estimated that a trillion gallons of water fell on the area over a four-day period, enough to cover the entire county in thirty-three inches of water if it accumulated without any drainage. A research team found that drownings accounted for 81 percent of the deaths.[145] Of those who succumbed to drowning, 37 percent died either in a vehicle or while leaving a vehicle. Indeed, tragedies like these have prompted some weather scientists to argue for deemphasizing Saffir-Simpson categories because they are tied entirely to wind; a category 1 storm might sound to a layperson like a relatively weak hurricane, but the wind categorization doesn't inform the public about the water danger.

Water carries an additional deadly dimension referred to as storm surge. When the winds of a hurricane sweep across the ocean or a gulf, they force the water in toward shore in a surge that can flood the adjacent areas. If a storm surge occurs during high tide, the combination can be especially calamitous, putting roads and homes underwater with perilous speed. When Hurricane Hugo struck South Carolina in September 1989 after pummeling the Leeward Islands and then Puerto Rico, the storm tide—that is, the storm surge plus the high-tide water—reached over twenty feet in some areas. The National Weather Service preserved an account of the sudden flooding by George Metts, a paramedic who had been stationed at Lincoln High School in McClellanville, South Carolina. The high school was being used as a shelter for area residents. Metts described how rushing water woke him at 1:38 a.m.[146] "We turned on our flashlights and could see water rushing through the air conditioners and water rising rapidly around our boots and equipment," Metts said, adding that he and a colleague struggled to open the door of the room they were in. "All this happened in less than five minutes," he said. He wound up spending several harrowing hours with hundreds of evacuees in the school cafeteria in chest-high waters, holding a three-year-old child above the surface while those who could huddled on tables. "The tidal surge had risen so rapidly that we had no time to call for help," Metts said. Fortunately, when the waters receded, no one in the room had been hurt.[147]

Wind. Rain. Surging seawater. When a major hurricane throws all of these at a populated area, the toll in damage and lives can be substantial. Despite the magnitude of disaster seen from storms like Katrina in New Orleans and Sandy along the Eastern Seaboard, the US mainland's geography and robust emergency infrastructure can help offset the danger. Islands usually have it worse, as when Hurricane Maria blasted Puerto Rico in 2017, wiping out homes and effectively destroying the power grid. Maria's death toll has been debated; a study published in 2018 in *The New England Journal of Medicine* suggested that at least 4,600 deaths occurred due to the hurricane and its aftereffects.[148] In less developed regions, the impact can be even more horrific. A November 2013 tropical cyclone known as Typhoon Haiyan ravaged the Philippines, taking at least 6,000 lives there.[149] More staggering still, Cyclone Nargis in 2008 is believed to have killed at least 138,000 people in Myanmar. That astonishing number partly reflects the dire impact of the storm surge in the low-lying Irrawaddy Delta as well as poor infrastructure and insufficient preparations. The country's military government then made matters worse by reportedly interfering with international aid efforts.

Some hurricane dangers can't be avoided. A tropical cyclone's winds will hit wherever the storm takes it. Other hazards are best addressed on long time frames, by ensuring homes and offices get built with hurricane threats in mind or even by defending entire cities. To protect New York City from the type of storm-surge flooding seen in Sandy, officials have discussed options including massive seawalls that could take decades to plan and construct. On a shorter lead time, individuals in vulnerable areas can conduct preparations ahead of each new hurricane season, ensuring emergency supplies of food and water are on hand.

But when an actual storm spins up over the Atlantic, that's where forecasting comes in, with computer-model guidance at the heart of it. When we can predict where a storm will travel and how severe it will be, we can determine who doesn't need to worry, who needs to ready their home by boarding up windows and bringing in the yard furniture, and, most important, who needs to get out of harm's way. Improving the accuracy of those predictions makes forecasts more trustworthy—crucial for encouraging people to act on them.

"There are decisions being made today on forecasts I would never have imagined when I was a student," the man at the podium said. On January 8, 2018, I was seated in Ballroom D of the Austin Convention Center in Texas, listening to speakers at the ninety-eighth annual meeting of the American Meteorological Society (AMS). This particular session had been set up to review the 2017 hurricane season, which had ended only weeks before. It was already considered a historic year thanks to three major storms that struck parts of the United States with tremendous force: Harvey in Texas, Irma in Florida, and Maria in Puerto Rico. The speaker at the moment was Louis Uccellini, who was then the director of the National Weather Service. And he was talking about the response by officials in Florida when guidance from the National Hurricane Center indicated that a tropical depression had the potential to develop into a hurricane that would threaten the Sunshine State. Governor Rick Scott was persuaded to declare a state of emergency on September 4, when the brewing future Hurricane Irma was still a thousand miles away—six days before its eventual Florida landfall.

Uccellini's statement captured my attention. I was just beginning the research on what would become this book and I had traveled to Austin for the weather world's big annual convention. The yearly AMS meeting is a dizzying week of events, with sessions devoted to esoteric research topics such as "Measurements of Radiation-Induced Condensational Growth of Cloud/Mist Droplets," some of which I struggled to follow. But this town-hall meeting—titled "The Devastating 2017 Hurricane Season: Opportunities, Challenges, and Future Directions for the Weather Enterprise"—was easy to understand.

What grabbed me about Uccellini's comment was not only its assertion of progress but also the way he chose to frame that progress: not just better forecasts, but forecasts that could earn enough confidence and trust for a state governor to stake important decisions on them. I had started learning more about meteorology because I wanted to answer a few simple questions: Where, exactly, do our forecasts come from? How good are they, and what can make them better? But Uccellini's talk helped me appreciate that those questions address only part of the story. Ultimately what matters is doing something about the weather.

"What we've been working towards," Uccellini tells me when I catch up with him again a few months later, "is connecting the forecast and warnings to decision-making." Example after example had demonstrated, he says, that the good forecast isn't enough. "We have to focus on the last mile."

When Uccellini took over the leadership of the National Weather Service in 2013, he made that connection between forecasts and decisions a cornerstone of the agency's path forward. By the time he retired on January 1, 2022, he had led an enormous shift in Weather Service culture, encouraging the 122 weather forecast offices to work more closely with officials and other emergency-preparedness stakeholders (as I saw firsthand observing local forecasting in State College, Pennsylvania). It isn't hard to see how proud Uccellini is of that progress. "We were actually preparing for Irma, preparing our interactions with emergency management, while it was still a wave off Africa," he says. "We've got to be able to do that more systematically."

Irma underscored both the power of preparation and its limits. This was an exceptionally strong and long-lived hurricane. It achieved category 5 status and made seven landfalls as it moved over Caribbean islands, the Florida Keys, and, finally, Florida itself. On Saturday, August 26, 2017, the National Hurricane Center—which at the time was still monitoring the dwindling remnants of Harvey over Texas—issued a brief advisory: "A tropical wave over western Africa is forecast to emerge over the far eastern Atlantic Ocean on Sunday." By the time Irma fully dissipated on September 13, the storm had claimed forty-seven lives as a direct result of its winds and rains.[150] Most were in the Caribbean; ten were in the United States. Indirect deaths—such as people who succumbed to heart attacks from the stress of the event and car-accident fatalities on rain-soaked roads—numbered eighty-two in the United States, far outpacing the direct deaths.

But the toll could have been far worse, experts later concluded. Working from the initial emergency declaration, officials encouraged evacuations well in advance of landfall. An estimated six million Florida residents fled from flood-prone coastal areas, according to the NHC. The early notice and the consistent messaging to the public about the threat, which were based on NWS and NHC forecast guidance, were credited with making the massive evacuation successful. If anything, some experts later speculated, the state had over-evacuated; some residents in areas without evacuation orders left anyway when they heard the forecasts and warnings. Irma's danger was most evident in the Keys. Monroe County, which includes nearly all of the Keys, put the number of homes destroyed there at over 1,100. Statewide in Florida, 64 percent of all customers lost power, one of the largest outages in US history. NOAA estimated the cost of Irma's damage in the United States at over $63 billion.

There are two key measures of hurricane forecast accuracy: track and intensity. Anyone who has watched hurricane coverage on TV understands the track forecast. Meteorologists chart the most likely path for the storm to take as it spins across the ocean. Accurate track forecasts are enormously helpful because they predict if, where, and when a tropical storm will make landfall. Intensity forecasts, which project the strength of a storm in terms of wind speed as it moves along the expected track, can be at least as important. Wind speed doesn't just affect people in the hurricane's path; it also factors into that dangerous storm surge. Some hurricanes undergo rapid intensification, when winds pick up by about thirty-five miles per hour or more within a twenty-four-hour period. Forecasting intensity—and, in particular, projecting rapid intensification—is generally more challenging than predicting tracks. Of course, insight into both where a hurricane will hit and how strong it will be at that point represents critical data for emergency officials making decisions about preparations and evacuations.

Uccellini stresses that accurate models aren't enough. "I keep coming back to this idea that it's not just the model forecast, it's how it's used that makes the difference. We've shifted our focus on the impacts of our forecast and our connections with decision-makers to help people understand these forecasts and make better decisions," he says. "Because if they're making the wrong decisions, even with great forecasts, then people are still going to die. This is what we're trying to avoid."

Five years before Irma, a soon-to-be-famous storm began its path across the Atlantic. We know it now as Superstorm Sandy. By Sunday, October 28, 2012, there was little doubt that this hurricane would make history.

Six days earlier, it had reached tropical-storm status and been assigned the name Sandy according to the protocol set out by the World Meteorological Organization. By Wednesday, October 24, Sandy's winds had picked up enough to warrant a hurricane designation by the National Hurricane Center. Sandy dragged northward across Jamaica and Cuba, then weakened as it headed up the Atlantic Coast, keeping east of Florida, to tropical storm status, but then it strengthened again to hurricane-force winds. Forecasts indicated that an unusual configuration of pressure systems would stop Sandy's northeast movement and push it back toward the coast and some of the most populated and vulnerable areas of the northeastern United States. This scenario implied catastrophic dangers.

That Sunday, in the Mount Holly, New Jersey, forecast office of the National Weather Service, Gary Szatkowski, then the meteorologist in charge, took an extraordinary step. He issued a PowerPoint-style message titled simply "Personal Plea." In the bullet points of this slide, Szatkowski asked that everyone follow evacuation instructions from state and local officials. "If you are still reluctant," he wrote, "think about your loved ones, think about the emergency responders who will be unable to reach you when you make the panicked phone call to be rescued, think about the rescue/recovery teams who will rescue you if you are injured or recover your remains if you do not survive." He described the coming storm as "extremely dangerous" and said that anyone who thought the storm "over-hyped and exaggerated" could call him later in the week to complain. He added: "I will listen to your concerns and comments, but I will tell you in advance, I will be very happy that you are alive & well, no matter how much you yell at me."

A detailed after-action service assessment from the National Weather Service released in May 2013 highlighted Szatkowski's exceptional plea for its "frank and persuasive language and effectiveness in focusing immediate attention on the gravity of the situation."[151] Surveys had found that emergency managers "pointed to this plea as the point when they realized the true magnitude of the threat."

As Monday, the day after Szatkowski's warning, dawned, Sandy had begun its dramatic swerve toward a head-on collision with the New Jersey coast. Around seven thirty p.m., according to the National Weather Service, the storm made landfall near Brigantine, New Jersey. The region was on full alert. Businesses closed and workers stayed home. Transportation officials had shut down the New York City subway along with buses and commuter trains on Sunday evening, a critical precaution given the level of flooding Sandy brought to the subway tunnels. When the storm passed, the United States had suffered seventy-two direct deaths and at least seventy-five indirect deaths, according to the service assessment. Damages were estimated at over $88 billion (adjusted for inflation in June 2024 dollars). The region would never be the same.

Sandy was historic for more than the depth of tragedy and loss. It also marked an inflection point in the evolution of weather forecasting in the United States. Reviews of the performances of various numerical models showed how prescient the ECMWF projections had been—and how the primary medium-range model for the United States, the Global Forecast System, had failed. As a National Hurricane Center report later described it, the ECMWF model predicted the swerve back toward the US coast at six and seven days in advance when other models had Sandy out at sea.[152] Those ECMWF track predictions are the forecasts that eventually wound up memorialized in artwork on the supercomputers in Reading. The aftermath of Sandy drew both public and official scrutiny of how the GFS performed against its European counterpart. When Congress passed the Disaster Relief Appropriations Act of 2013 to provide supplemental funds for the Sandy recovery, it included money for NOAA to upgrade computers and improve the GFS.

A few years later, I had a chance to see this work in action. After taking exit 9 off the New Jersey Turnpike and heading south on US 1, I arrived at the Geophysical Fluid Dynamics Laboratory (GFDL), a NOAA research facility. On the day I visited in 2018, a big GFDL project aimed at improving numerical weather prediction was on the cusp of implementation.

I was there to meet Shian-Jiann Lin, a veteran GFDL scientist who had led the work on this project. Like so many of the people I met across meteorology research, Lin—usually called "SJ" by his colleagues—can trace his interest in weather back to childhood. Lin was born in 1958 on Taiwan, an island notorious for the annual tropical typhoon season. Lin, who has since retired, remembers being fascinated by the physics of the monster storms. What drove the movement of the storms? What fed their destructive energy? "The more time you spend looking at the weather, the more you realize it's really part of your life," he says.

Lin's work at GFDL was a new and radically different dynamical core for the GFS. The dynamical core is the component of the model that serves as its engine. It holds a representation of the atmosphere and performs the calculations that describe how key attributes such as temperature and pressure will change moving forward in time as parts of the atmosphere interact with each other. The GFS had long relied on a spectral dynamical core, in which the atmosphere is depicted and processed as big wave-like patterns interacting with each other. Lin had devoted years of work to developing an alternative approach: the finite-volume cubed sphere, usually referred to as FV3. In contrast with the waves of the spectral method, FV3 divides the atmosphere into boxes. Each box influences the surrounding boxes in ways that can be calculated with mathematical formulas. Both the spectral and finite-volume approaches have their pros and cons and there's no simple answer as to which is superior. There are too many variables, including the design of other parts of the model and how best to use computing power. The ECMWF produces its winning results with a spectral core. But for the GFS, NOAA's testing found accuracy improvements with FV3.

As I watch, Lin types a command on his desktop computer: *msub C_1536_0801_e76*. The command travels from Princeton to a supercomputer system named Gaea (after the Greek goddess of the earth) in Oak Ridge, Tennessee, to launch a run of the experimental FV3-based model. It takes eighty minutes to complete the full ten-day run. When the results appear, Lin examines them as a map. "This is hot," he tells me. "This is pretty hot." He's talking about the temperatures a week in the future, around the Fourth of July. Sure enough, that Independence Day saw a heat wave sweep large parts of the country.

A year after my visit to GFDL and nearly seven years after Sandy, the operational version of the GFS used by National Weather Service forecasters across the country switched over to the new FV3 core. The resulting experience with the improved GFS underscores both progress and challenges in advancing US numerical weather prediction. Analyses of the forecast skill of the GFS indicate that its predictions are getting better. But the competition keeps improving too. Most experts continue to view the ECMWF's model as the world's best at medium-range forecasting. When it comes to severe weather or key national forecasts, meteorologists in the United States can—and will—look at the GFS, the ECMWF, and other models. But for routine, everyday forecasts, including some automated predictions, the GFS serves as a workhorse for the National Weather Service. And critics have charged that NOAA simply isn't agile enough when it comes to advancing its weather modeling.

It wasn't until after I spent time at the ECMWF that I fully appreciated some of the contrasts. But it's not hard to understand. All you really need is a map. Look at where NWP work takes place in the United States and you'll cover a lot of territory. In Princeton, there are the scientists at GFDL devoted to research. The operational models are supervised by a different organization: the National Centers for Environmental Prediction in College Park, Maryland. Another research group, the Environmental Modeling Center, also makes its home in College Park. More research takes place at the National Center for Atmospheric Research, a federally funded facility in Boulder, Colorado. Then there are the top research universities across the United States. To be sure, each of these places contributes knowledge and brainpower; the sheer size of US weather research is a great strength.

But that size comes with drawbacks too. Cliff Mass, professor of atmospheric sciences at the University of Washington, is one of the best-known—and most outspoken—critics of US modeling. All the way back in 2006 he published a paper in the *Bulletin of the American Meteorological Society* titled "The Uncoordinated Giant."[153] For the United States to fully realize its potential, he argued, required tighter ties between researchers and their operational counterparts. After all, if you wanted to produce the best automobile or airplane possible, you'd want the designers and engineers communicating closely with the people who build and test the products. In 2018, when I spoke with Mass ahead of the FV3 transition, he contrasted the diffuse US work with the tight focus of the ECMWF. "In the end you have to find a way to bring everybody together," Mass told me.

To try to make that happen, there's an ambitious US effort underway. It seeks to take a page from Silicon Valley and the move-fast ethos of tech companies by tapping the power of open-source development and cloud computing. The program is called the Unified Forecast System (UFS), described as a "community-based" modeling system, with the community in question consisting of researchers and developers throughout the weather enterprise, not only those inside NOAA's modeling teams. Work on the UFS is done by NOAA, through a new organization called the Earth Prediction Innovation Center, universities, and the private sector as well. One goal, as laid out in a 2022 paper, is to streamline twenty-one different operational weather models run by NOAA into eight applications built from the same UFS framework.[154] For instance, NOAA has depended on the GFS for medium-range forecasts and a different model, the High-Resolution Rapid Refresh, for short-range high-resolution predictions, but the new framework would use the same underlying technology for both. (I'm reminded of the television chef Alton Brown, who rails against "unitasker" gadgets devoted to a single purpose in the kitchen, such as egg slicers, when the job can easily be handled by a good knife. More to the point, the new approach is similar to the ECMWF's, which uses one model to handle different resolutions.)

"Why can't we all work together?" For Neil Jacobs, who has been a driving force behind the UFS effort, that question helps illuminate the lagging pace of US modeling. "If we all work together and pool our resources and pool our brainpower, we'll have an A-plus model." Under the new approach, anyone can download the computer source code for the UFS model from GitHub, a popular home for open-source software development, allowing developers anywhere to play around with it and propose their own improvements.

Neil Jacobs now serves as chief science adviser for the UFS. He brings an unusual background to the role; he's a weather-modeling insider yet mostly a NOAA outsider. His government experience came during the Trump administration when he was an assistant secretary of the Commerce Department, the cabinet-level parent of NOAA. After the administration struggled to get its pick for NOAA administrator confirmed, Jacobs wound up as acting NOAA administrator until the incoming Biden team named its own choice in 2021.[155]

Jacobs earned a Ph.D. in atmospheric sciences with a focus on numerical modeling from North Carolina State University and joined the weather world's private sector at AirDat, a company that was developing a way for commercial aircraft to gather valuable atmospheric observations as they plied their routes.[156] AirDat wanted to be able to run its own copy of the GFS model to understand how plugging in its own data would work. In his private-sector role, Jacobs says, he learned a lot about how finicky technical requirements effectively walled off the GFS from the non-NOAA world, limiting the ability of outside researchers and even some within NOAA itself to experiment. During his relatively short stint in government, Jacobs collaborated with William Lapenta, a respected

NOAA veteran, to devise a way to open up weather-modeling work—an effort that birthed the Earth Prediction Innovation Center. (Lapenta died in 2019 in a drowning attributed to rough surf off a North Carolina beach.)

Along the way, Jacobs was swept up in what became known as Sharpiegate, an absurd-sounding yet nonetheless serious incident that highlighted the perils of politicians recklessly interfering with scientific experts.[157] In Sharpiegate, President Donald Trump created confusion about the forecast for Hurricane Dorian in 2019, first with a Twitter statement and then by displaying a map that appeared to have been altered with a Sharpie marker, Trump's well-known writing instrument of choice. National Weather Service meteorologists rushed to reassure people in Alabama that they were not, despite the president's comments, under threat from the storm. Reporting by *The New York Times* found that Trump, angered at being contradicted, had an aide contact Jacobs's boss, secretary of commerce Wilbur Ross, a Republican financier appointed to the cabinet post by Trump.[158] Ross in turn went to Jacobs. As *The Washington Post* described it: "That led to an unusual, unsigned statement from NOAA [. . .] that backed Trump's false claim about Alabama and admonished the National Weather Service's Birmingham, Alabama, division for speaking 'in absolute terms' that there would not be 'any' impacts from Dorian in the state."[159]

The episode raised serious concerns about political pressure influencing the hard science and professional opinions of Weather Service experts. A 2020 report from the Commerce Department's inspector general faulted Jacobs for having "acquiesced" but also made clear the extent of pressure from above.[160] In the days after the controversial unsigned statement, Jacobs took pains, first at an industry conference and then with an all-staff email, to praise the Alabama NWS team that had worked to correct Trump's inaccurate information.

When I spoke with him, Jacobs appeared much more comfortable in his latest role at the nonprofit University Corporation for Atmospheric Research, advocating for a better and faster approach to evolving the US weather models. Like others I talked to, he sees the kind of collaboration that the ECMWF fosters among its scientists as a necessary component for the United States to up its modeling game. Already the Earth Prediction

Innovation Center has disparate parts of the US weather enterprise talking more with one another. Still, transitioning NOAA operations to the eventual unified models is expected to take years—but Jacobs argues the payoff will be worth it, given growing dangers from weather threats.

"The risk exposure is growing exponentially," he says, with both climate-driven extreme events and population-development growth that puts more people in risky areas. "And so, in order to keep up with that, we need better forecasts." He adds that ultimately those kinds of results drive weather scientists in their work. "There's not a lot of people who can go home at the end of the workday and say, 'I prevented somebody from dying.' When you talk to these forecasters, they don't make a lot of money. They're not there for the money, they're there for the mission."

Numerical weather prediction represents some of the most mathematically sophisticated and computationally intensive work that you can imagine. Conceptually, though, the idea is simple. If you can simulate the behavior of Earth's atmosphere, you can step forward in time and gain a picture of what the weather will look like in the future.

Simulations are all around us in the computer age. Airline pilots train in cockpits that never leave the ground but nonetheless display computer-generated images out the window and instrument-panel readings that faithfully show the results of flipping a switch or nudging the controls in a particular way. Surgeons can practice complicated procedures on a virtual patient designed to respond the same way a real human body would. You may have played around with simulations yourself. In the computer game *The Sims*, you control a character that interacts with its world, working in a job to earn money and meeting other Sims to form relationships.

What all of these simulations have in common, regardless of their complexity, is a model of an environment and rules that govern the model's behavior. In *The Sims*, for instance, a rule says that if you haven't instructed your Sim to eat, it will become hungry and eventually die of starvation. The verisimilitude of any simulation depends on how faithfully the model hews to reality and how well its rules reflect the consequences of decisions and changes in the environment.

Modeling the weather presents an enormous challenge. The environment being modeled is the entire atmosphere of the planet. The rules reflect the complex behavior of the gases that make up that atmosphere. Variations in pressure and temperature at different altitudes, warming from the sun, the amount of water vapor—all these and more affect the way those gases move around, remaining relatively calm at some points and creating wind and storms at others.

The foundations of numerical weather prediction date back to the early twentieth century. Cleveland Abbe, an American meteorologist who had spearheaded the establishment of weather-observation networks, proposed the use of mathematics and the application of physical equations to forecasting.[161] (Abbe was the first chief meteorologist of the US Weather Bureau, the predecessor of today's National Weather Service.) On the other side of the Atlantic, a Norwegian scientist by the name of Vilhelm Bjerknes similarly posited what would become the basic concepts of modeling: that by taking the current state of the atmosphere and performing calculations that govern thermodynamics (the physics of heat), hydrodynamics (the physics of how fluids behave), and other forces at work, it would be possible to predict the future state.[162] A few years later, English scientist Lewis Fry Richardson attempted just such a forecast, working through the equations by hand.[163]

But even if the concepts were sound, these early pioneers were limited on multiple fronts, from the lack of detailed observational data needed to describe the initial state to the labor of solving the necessary equations. Computing pioneer John von Neumann—a key scientist on the Manhattan Project—championed meteorology as an application for ENIAC, considered the first true digital computer. Eventually, the rise of digital computing, weather satellites, and other data sources provided solutions to those challenges. By the 1970s, supercomputers crunching the data from satellites and radar marked the arrival of serious operational NWP.

Today, whether at the ECMWF, NOAA, or elsewhere, NWP relies on some key components, such as observations, the dynamical core, the physics package, and data assimilation. Observations provide the information that describes the real-world state of the atmosphere: temperature, pressure, winds, moisture, clouds, and more. The dynamical core is that engine where the key calculations take place. The physics package refers to a set of rules that fine-tune the operation of the dynamical core. Data assimilation is how new observations get incorporated into the model as it runs.

Of course, it helps to get better observations. Putting a more accurate and detailed view of the atmosphere into an NWP model improves the quality of the output. One way to do that: Build one of the most sophisticated weather instruments ever, then shoot it into the sky.

I am standing in one of the most uncommon rooms I've ever seen, looking at a foil-covered, antenna-studded object the size of a small school bus. Fifteen months later, this object will be 22,236 miles above me, watching over me and a significant swath of the Western Hemisphere.

The object before me is the US weather satellite designated GOES-U.[164] Along with its three sibling satellites already orbiting the Earth, GOES-U represents an essential tool for seeing the atmosphere as it is and as it will be. These satellites give us images that show the weather from a celestial perspective. They see the gigantic whorls of a hurricane, the snow-laden clouds of a nor'easter, even dust blown off the Sahara Desert's sands and lofted into the atmosphere where it can block sunlight and alter the weather. They are the indispensable eyes of modern forecasting, which means their vision must be faultless.

And that explains the unusual location I'm visiting in Littleton, Colorado. This is a clean room at Lockheed Martin Space, the defense and aerospace giant that built this latest generation of weather satellites. A clean room, as the name suggests, is a facility kept free of dust and all other contaminants. Spacecraft require these meticulous precautions, lest a tiny particle wind up fouling their circuits or sensors in orbit. The Geostationary Operational Environmental Satellites craft are particularly sensitive because they include optical sensors. Lockheed can't afford a mote in the spacecraft's eye.

I've seen clean rooms in photographs and space-program documentaries, but those haven't prepared me for the experience of actually entering one. As I gaze at GOES-U, every part of my body is covered except for about seven square inches of my face, a narrow strip exposing my eyes and the bridge of my nose. It took fifteen minutes and the expert tutelage of Matt Kettering, a Lockheed engineer and experienced cleanroom hand, to get me suited up to requirements. The process involved, among other things, stepping onto a large sticky mat akin to a glue trap that pulled dirt from my shoes and donning a hair cover, mask, jumpsuit, hood, gloves, and booties. Looking down at my white-suited figure, I felt like one of the characters in *Willy Wonka and the Chocolate Factory*, ready for the Wonkavision room.

Only then was I considered ready to step through the doors into the clean room. The ventilation system inside not only scrubs any stray dust from the air but also keeps the pressure in the clean room slightly higher so that dirtier air from outside can't waft in. The standards are so exacting that I'm barred from bringing my reporter's notebook with me, since the paper might shed particles. Kettering supplies me with a blue sheet of special clean-room paper, designed to avoid contamination.

This painstaking process suggests how high the stakes are when it comes to these incredibly complicated spacecraft. GOES-U, the satellite in front of me on the clean room's floor, is the last of four craft in the GOES-R program, named for the first in the series. Official budget numbers have put the lifetime cost for the GOES-R program at over ten billion dollars. The development timeline for major satellites like this is measured in decades. Indeed, planning for the GOES-R series dates back more than twenty years. The first craft was launched in 2016. It was followed by the second launch in 2018, the third in 2022, and finally GOES-U in 2024. These satellites will serve through 2036, according to NASA and NOAA. By that point—at least under current plans—the succeeding generation will be in orbit, ready to take over.[165]

In other words, a great deal rides on GOES-U and its siblings working flawlessly twenty-four hours a day, seven days a week, three hundred sixty-five days a year for up to two decades. If problems develop in orbit, NASA and NOAA can't just whistle up a quick replacement. The program does include some redundancy: Daily operations are handled by two GOES birds, one positioned to focus on the eastern part of the United States and another for the western side. The others orbit on standby and can swap in if needed.[166]

When you think of weather satellites, your mind might first turn to pretty pictures, like a photo taken by the astronauts of Apollo 17 that became known as *The Blue Marble*. It captured the full disk of a sunlit Earth, revealing the sapphire of the oceans, the whites of clouds and the Antarctic ice cap, and the greens and browns of land. Weather satellites operate closer to the planet with sensors capable of much higher resolution. The first spacecraft in this latest GOES series immediately began beaming back lush images of our world. Its photographs display remarkable detail, recording plumes of smoke rising above wildfires and views into the hearts of hurricanes showing the churning eye walls. But for meteorology, the satellites' ability to see things invisible to the human eye provides some of the most important data. For instance, a key GOES sensor monitors four near-infrared channels and ten infrared channels, including a specific wavelength used to observe "upper-level tropospheric water vapor" that provides information about jet streams and turbulence. Other infrared wavelengths help detect the temperatures at the Earth's surface and how high the tops of clouds reach into the sky—all fodder for the NWP models working to simulate the atmosphere. Another instrument, the Geostationary Lightning Mapper, aids in detecting tornadoes and thunderstorms (for more on tornadoes, see chapter 1).

It's a long way from the early efforts to monitor the weather from above. When the space race kicked off in 1957 with the Soviet launch of the first Sputnik, simply getting an object to circle the Earth marked a huge achievement. In 1960 NASA lofted into orbit TIROS-1, considered the first true weather satellite. Now weather satellites are a fundamental source of information for modern meteorology. The GOES satellites, culminating in the craft I saw being prepped in Colorado, will be workhorses for NOAA for years to come. Meteorology draws on data from other satellites, including some that trace different orbits as well as craft launched and operated by other countries. Meanwhile NASA and NOAA are developing plans for the next generation of geostationary satellites to take over for the GOES craft in the 2030s, with improved sensors to gather even more data.

Until then, the current GOES satellites will watch tirelessly from their orbits, beaming back observations of the atmosphere and feeding them into the global models. That's where science, math, and technology combine to turn all that data into predictions.

"We have one model. We don't have a thousand. We have one," Florence Rabier tells me as we sit in her office in the ECMWF's Reading headquarters. As director general of the ECMWF, Rabier holds one of the most prestigious posts in global meteorology. She is also one of the world's foremost experts on numerical weather prediction. The government of her native France bestowed upon her membership in the country's highest order of merit, naming her a chevalier de la Légion d'Honneur—a knight of the Legion of Honor—in recognition of her contributions.

For all that, Rabier comes off as approachable and unhurried; during my visit, it was noticeable how comfortable ECMWF staffers were chatting with her in the hallways and the canteen. Before we sat down to talk, she gave me a brief tour around her office and showed me some of the mementos that underscore the global reach of the ECMWF, among them a photograph of Rabier at a European weather-satellite launch in French Guiana, a challenge coin given to her by Rick Spinrad, the administrator of NOAA, and, on top of a credenza, an elaborate clock made by Jaeger-LeCoultre; Rabier told me that it was an Atmos clock, powered entirely by the effect of slight temperature changes on the pressure of gas. The clock was a gift from the Swiss government to the ECMWF a few years after its establishment.

Rabier's remark about the ECMWF having one model comes as she explains why she believes the center has been so successful at improving forecasts. One factor is something I've heard repeatedly as a point of pride among scientists and staff: The ECMWF draws from the top scientists and researchers across its member and cooperating nations, benefiting from the diversity. It also pays well enough to attract and keep this talent. But the ECMWF also benefits from a distinctive clarity of focus. The entire organization gears itself around the operation and advancement of that one model Rabier is talking about, colloquially known (particularly in the United States) as the Euro but officially named the Integrated Forecasting System, or IFS.

To fully appreciate Rabier's point, it helps to reiterate some of the things the ECMWF does *not* do. It doesn't produce weather forecasts for the public; it supplies the output from the IFS to the individual weather agencies at the member nations while providing training to help those agencies get the most from the model forecasts. The ECMWF doesn't collect local weather measurements like the daily highs and lows charted at each NWS forecast office. It doesn't issue watches or warnings. People here obsess over how to improve the IFS.

"We all have the same goal," Rabier says. "Optimizing the model. Giving the best forecast. Everybody works towards that, so I think it focuses the mind." She argues that the independence of the center allows it to move nimbly when it comes to implementing upgrades that improve the model. "The member states keep us on track. That's what we want. But they don't interfere on a day-to-day basis. So if we think our next upgrade is good, we implement it. We don't have to ask for the opinion of everyone." Research and operations work under the same organization, making it easier to ensure that cutting-edge advancements get incorporated into the live IFS as soon as they've proven themselves. That research-to-operations pathway is something that critics of NOAA have complained about going back to Hurricane Sandy and earlier, though the agency has made changes in recent years to speed improvements.

Florian Pappenberger, ECMWF's deputy director general and director of forecasts, also emphasizes the power of autonomy. "Our decision process of implementing an upgrade cycle is very simple," says Pappenberger. "The director of research decides to hand it over to me, and I decide that it gets implemented." It's a lot of responsibility, Pappenberger acknowledges, but the streamlined process avoids bureaucratic delays.

The energy of the place feels less like a multinational agency and more like a cross between a university and a corporate research center with some distinctive quirks sprinkled in. Take, for instance, the rubber duckies. Outside the main entrance in Reading stands a minimalist-style fountain that employees took to decorating with rubber ducks. Over the years, the duck fountain has become an unofficial symbol of the center. Guests often bring ducks to add to the collection; I spotted ones with the logos of tech giants IBM and Juniper Networks among the dozens scattered across the fountain space.

Lunch at the staff canteen feels like a conference mixer, with conversations about snowfall and surface temperature going on while scientists from different specialties line up for enormous helpings of spaghetti Bolognese or a healthier fish option. The topic of climate change pervades many of these discussions as inescapable context and a hurdle to be met with better forecasts. "There will be more extreme events," Rabier tells me. "So we need to predict them well. Extreme events mean high resolution." During my visit, the center was in the midst of increasing the resolution of IFS medium-range forecasts to nine kilometers from eighteen across all of the ensemble members, an upgrade that demonstrated better accuracy for both the track and intensity of tropical storms. The ECMWF also has responsibilities related to multiple European Union climate initiatives, including the Copernicus Climate Change Service, a program for data and forecasts to develop and assess specific climate solutions, and Destination Earth, a so-called digital-twin simulation of the planet to explore climate scenarios.

For an observer from the United States—or this observer, anyway—the extent to which climate change comes up in conversation is noticeable. I was surprised when Rabier and several other colleagues independently mentioned the need to manage the carbon footprint of the energy-hungry supercomputers that solve the NWP equations. US weather officials I've spoken with are keenly aware of the impact of climate change, but I couldn't recall a single instance of someone talking about the need to reduce the emissions from their work. (It's worth noting that Europeans are much more likely to describe climate change as a major threat than Americans are, according to a 2022 poll by Pew Research Center.)

The ECMWF traces its establishment back to the 1960s, when European nations began to focus on pooling their resources and knowledge in the sciences. The formal document establishing the ECMWF was signed by the initial group of member nations in 1973; it proclaimed the "importance for the European economy of a considerable improvement in medium-range weather forecasts" and stated that "the improvement of medium-range weather forecasts will contribute to the protection and safety of the population." The center was set up near the University of Reading, which has a respected meteorology program.

Five decades later, the ECMWF's forecast model and its culture of steady improvement and innovation are admired throughout the weather community. Participation expanded in the twenty-first century to include a spate of additional countries, among them former Soviet republic Estonia, a full member, and Israel and Morocco, cooperating states. Brexit created a thorny situation, and the ECMWF has since opened two facilities outside the United Kingdom: an office in Bonn, Germany, and a data center in Bologna, Italy, with a new generation of supercomputers to provide an upgrade over the already formidable ones here in Shinfield Park with the Sandy forecast stenciled onto their cabinets. Even so, in 2021 the ECMWF committed to keep its headquarters in Reading and has plans to construct new offices on the University of Reading campus.

The tight focus, relatively compact size, and day-to-day autonomy of the ECMWF—so different from the immensity of NOAA and the National Weather Service—pay off. It's difficult to overstate the influence of the ECMWF, both for its achievements so far and for the expectation of future progress based on its track record.

Near the end of my visit, I had an unusual opportunity to see the ECMWF's work and culture in action. Every Friday afternoon, staffers come together for an all-hands event known simply as the weather discussion or the weather meeting. At this gathering, meteorologists review selected forecast developments from the week, examining anomalies in the IFS model's performance and debating if and how an issue needs to be resolved. The weather meeting is not an event for outsiders. The ECMWF doesn't typically allow journalists to attend; its leaders want to ensure an open discussion where everyone feels able to speak candidly, questioning their colleagues and offering constructive criticism. I've agreed to refrain from quoting specifics from the discussion and limit my observations to general impressions, which isn't much of a burden, given how technical some of the conversation turns out to be.

The meeting takes place in the Weather Room, a conference space off the center's main lobby, a few steps from the cafeteria. Screens cover the wall at the front of the room, controllable by a touch-screen panel mounted on a stand. Along one side of the Weather Room, glass walls enclose a couple of computer workstations. It is the center's practice to have a forecaster on duty inside that fishbowl, keeping track of the model's prognostications and how well they pan out (*verify* in NWP parlance) and writing a daily report. I've spotted one person in that fishbowl several times during my visit, looking quite busy every time. He turns out to be Tim Hewson, who leads the forecast-performance monitoring and product-development teams. He serves as a critical link between the researchers and developers improving the model and the people who actually use the forecasts. "At the end of the week, we take the culmination of all you've done in the week, and that gets presented on a Friday afternoon in the weather discussion," says Hewson, who was chief forecaster at the UK Met Office before moving to the ECMWF. "The idea is that people can come here and chat [about] what we've been looking at or what the analyst on duty has been looking at and help to address and resolve issues and bring ideas onto the table."

Around two thirty p.m., people begin filing into the Weather Room, some moving chairs around to sit together, others leaning against low bookcases at the edge of the space. There are about forty people present in person and another few dozen attending remotely from Bonn. Hewson begins with a report on a bug in a coming update that has been forecasting improbable amounts of snow during testing. Standing before the video wall at the front of the room in blue jeans, an untucked shirt, and running shoes, Hewson shows some slides as he outlines the status of the problem and the efforts to fix it. After he finishes the overview, the tone of the gathering shifts from meeting presentation to fast-moving conversation, with audience members musing aloud about options and firing off questions that others respond to. Suffice it to say that the discourse is acronym-laden and features an impressive variety of accents, reflecting the international makeup of the staff.

After a spirited exchange about the snow bug, Hewson moves on. A late-February winter storm prompted nearly unprecedented blizzard warnings in Southern California, where residents marveled at the sight of snow falling around the Hollywood sign and at Disneyland. (Technically, the white stuff may not have been snow but the wet frozen pellets known as graupel.) Higher elevations were inundated by eight feet or more of snow. Hewson notes that the ECMWF's models had forecast the snow seven days in advance, a win he described as "good early warning." The weather discussion wraps up on a lighter note with a traditional meeting ender: "Weather for the Weekend." Hewson tells the group the forecast for the Reading area calls for colder temperatures. "Rather cloudy, rather cold," he says. Naturally, he turns out to be correct.

Once the supercomputers at the ECMWF and the other global-modeling centers have finished their calculations and kicked out their predictions, doing something about the weather requires turning the output from the models into actual forecasts, including the watches and warnings that put officials, communities, and anyone in a storm's path on alert. For that, the action moves four thousand miles away from Reading to a building on the Miami campus of Florida International University: NOAA's National Hurricane Center.

The distinction between model projections and forecasts has grown in importance in the age of ubiquitous internet access and social media chatter. Now anyone can call up a web page showing the so-called spaghetti plots for a hurricane. These are charts with a dozen or more squiggly lines superimposed on a map. On the typical hurricane spaghetti plot, each line indicates the projected track of the storm according to a specific model. The same approach can be used to show different ensemble members from one model. In the hands of a knowledgeable forecaster, the spaghetti plot offers a useful tool to quickly ascertain whether different models expect the storm to follow a similar track or whether the models disagree wildly.

But for non-meteorologists, this kind of information can make it all too tempting to look at the plots and draw a conclusion that supports a preference. *Well, I don't need to worry about that hurricane watch because several of the models say the storm will make a hard turn before it gets to me,* one might conclude. Worse, social media personalities may play up disagreements among the models to help generate clicks. The easy access to NWP model information overlooks the reality: Models are a tool for meteorologists to predict the behavior of dangerous storms, not a choose-your-own-adventure game for armchair weather buffs. The scientists at the ECMWF would be the first to agree with this. Florence Rabier and her colleagues focus on running and improving their model, leaving the issuance of forecasts to the meteorology agencies at member nations. In the

United States, the global model gets handled by other parts of NOAA. The NHC meteorologists analyze all the key models as they build a forecast.

"There is a lot of forecaster experience and intuition that goes into the forecast," says Dan Brown, branch chief of the Hurricane Specialist Unit at the NHC. Brown joined the NHC in 1993 and now oversees the elite team of NHC meteorologists who issue forecasts—and, more critically, warnings—for tropical storms and hurricanes. Though Atlantic storms get most of the attention, the NHC also handles forecasts for the eastern Pacific, including Hawaii. And it isn't just the United States that depends on the NHC; countries throughout the Caribbean and Central America take recommendations from the center regarding tropical storm and hurricane warnings. (Often the NHC forecasters are predicting their own weather, given their headquarters' location in Miami.)

When Brown talks about experience and intuition, he nods to the fact that different models often disagree. And the NHC looks constantly at a whole bunch of models. The ECMWF model and NOAA's GFS may be the most prominent global models but NHC forecasters examine others as well, including the model operated by the UK Met Office and another from the US Navy. Then there are more specialized models optimized for tropical storms, such as the Hurricane Weather Research and Forecasting Model (abbreviated as HWRF and pronounced "H-wharf"). The specialized models typically zoom in on a defined region rather than the entire globe and can provide much finer geographic resolution. Beyond the models, NHC forecasters—like most operational meteorologists—look at observations, including satellite imagery, water temperatures from ocean buoys, and radar data. On top of that, they can also dispatch the aircraft teams known as the Hurricane Hunters, who fly specially equipped WP-3D Orion turboprop airplanes to gather radar and other readings from inside the storms—not a mission for the faint of heart. In other words, while some TikTok amateur nods sagely at the latest spaghetti plots, Dan Brown and his colleagues digest a staggering amount of information from different sources to inform their forecasts.

It works. Time and again, if you look at a chart showing hurricane forecast accuracy, the NHC's official forecast outperforms any specific model. Each year, NHC produces a forecast-verification report for the previous tropical-storm season. Meteorologists focus on two key aspects: the track and the intensity. According to the NHC's 2023 report describing the 2022 Atlantic season: "Records for track accuracy were set at 24, 36, 48, 60, 96 and 120 [hours]."[167] The report also noted that official track forecasts were "slightly outperformed by consensus models at most time periods," a fact that further underscores the pitfalls of following any specific model too closely—*consensus models* refers to models that draw on projections from multiple different models. (This differs from ensembles, where multiple forecasts are generated by the same model but with slight variations introduced.)

But the most important elements of the official NHC forecasts are the watch and warning alerts critical for understanding potential impacts and dangers from a storm. "We really don't want someone going to look for the models," Brown tells me. "We don't want someone worrying about what category it's going to be. We really want folks to understand if they're under a storm surge or hurricane watch or warning. We want folks to listen to their local officials. And we really want folks to be focusing on those hazards." Brown adds that sometimes people who are in the path of a storm focus on one particular model track or on whether the storm is expected to reach a particular category on the Saffir-Simpson scale. It's the forecast world's equivalent of people listening to WebMD rather than their doctors. "People will say they're not going to move unless it's a category three or four or five," Brown says. "But category ones and twos produce a lot of damage and can be significant as well, especially when it comes to water, the storm surge, and rainfall."

A key asset that human forecasters bring to this world of number-crunched models can be summed up in one word: *consistency*. "The models can flip-flop back and forth," Brown explains. "But we as forecasters strive to provide a fairly consistent message in our forecasts. When we make changes, we make fairly deliberate, fairly steady changes." In essence, they use their real-world experience, along with all the other observations available, to smooth out changes from one NWP model run to the next, ensuring that no single forecast unduly overreacts to a computer-projected storm swerve. "It helps build credibility and trust," Brown says. If people see too many dramatic swings in forecasts, they may be less inclined to take any one of them seriously.

Human forecasters at the NHC also shoulder the work of communicating what the expected path and intensity of a hurricane means in the real world. They must help translate the meteorology into the information that people—both officials and the public—need to make the best decisions. That means focusing on impact and risk, issuing watches and warnings. A National Weather Service hurricane watch gets issued when forecasters expect that a storm with winds of 74 miles per hour or more "poses a *possible* [italics mine] threat, generally within 48 hours," according to the official definition. A hurricane warning means winds of that level are "*expected* in 36 hours or less." [Again, italics mine.] Tropical-storm watches and warnings operate the same way but for storms where the winds aren't expected to reach hurricane levels. In 2017, the Weather Service rolled out two additional types of guidance to focus on the perils of flooding. It bears repeating once more: Water kills. So forecasters can now issue a storm-surge warning when there is "a danger of life-threatening inundation from rising water moving inland from the shoreline." A storm-surge watch indicates potential for such conditions that hasn't yet risen to the level of danger.

"We really hope folks are using these as a tool to understand that when you're under a storm-surge or hurricane watch or warning that it is time to either think about or start taking action," Brown says. As always with weather, there's no guarantee the forecast will play out as expected. "But it means the risk is high enough that you need to take action to protect your life and your family."

In recent years, NOAA has stepped up its efforts to improve hurricane forecast communications through findings from social science. Similar to social science projects for tornado messaging and other weather threats, this work involves surveying relevant groups, from emergency managers to members of the public, for insights into how they understand and use communications from NHC and local forecast offices. Among other things, this work has helped prompt the use of blunt and direct language to ensure critical information doesn't get lost in forecast jargon, just as Gary Szatkowski sought to galvanize people in Sandy's path. When Hurricane Laura swept across the Caribbean in 2020 with a track that forecasters projected would slam into the Louisiana coastline with terrible force, the NHC estimated the potential storm surge could send water as high as fifteen feet or more cascading over some areas.

"We're trying to pick and choose our words very carefully, very meaningfully, so that we can try to get the best response," Brown says. "That's when we came up with the word *unsurvivable*." As Laura closed in on the coastline near the Texas-Louisiana border, the NHC issued this message: "Unsurvivable storm surge with large and destructive waves will cause catastrophic damage." In an after-action report on Laura, the NHC described that language as "unprecedented."[168] It went on to note that "evacuation compliance in the hardest-hit areas of Cameron Parish is estimated to have been at or near 100% since there were no rescues after the storm, and as of this writing, there are no known deaths from Laura as a result of storm surge." There were, however, fatalities from other causes. The after-action report tallied seven direct deaths in the United States, including people killed by falling trees or drowning in surf. The number of indirect deaths—a category that includes people who died from carbon monoxide poisoning from use of portable generators, heat-related illnesses, and other factors—was thirty-four, which eclipsed the direct fatalities. Outside the United States, thirty-one people drowned in Haiti, and there were nine drowning deaths in the Dominican Republic.

I asked Brown about the implications of more accurate forecasts. Improvements in the NWP models are giving meteorologists the ability to make good hurricane predictions further out into the future. Take, for instance, the NHC's tropical weather outlook, which gives an overview of potential tropical-cyclone formation. This outlook had long been a two-day forecast. In 2013, NHC expanded the time period to five days. Then in 2023, it grew again to seven days. Forecasts of specific hurricanes currently project five days out, and Brown expects that will grow as well. But he says longer forecast horizons come with caveats, at least for those who don't appreciate the inherent uncertainties.

"I think people are making a lot of preparedness decisions far out in time. That's good, and that's bad," he says. "It's good that they're doing it. But they have to remember the forecast can change." He says social science research has found that people can get "anchored" on a specific forecast, which is particularly risky if you conclude where you live won't be affected. "I don't want people to get *too* confident. We don't want people to get so confident that they're not making preparations."

As Brown suggests, getting people to comprehend the dangers can be difficult. For the weather world's professional communicators, it's also an opportunity for innovation to make sure the message gets through.

The image on the television screen looks like a routine weather segment. Jim Cantore, a veteran Weather Channel meteorologist, stands on a sunny South Florida street. Nearby palm trees cast gently swaying shadows. In the background, waters from an Atlantic Ocean inlet lap at a seawall as Cantore talks with a man beside him, Manuel Bojorquez, a CBS News correspondent based in Miami, about climate change and hurricanes. "It's that warm water that fuels hurricanes. That water is getting warmer," Cantore says. "That helps to create more intense storms."

Suddenly the skies darken. Heavy rain cascades from above as wind whips the palm branches. As Cantore and Bojorquez continue to discuss hurricane threats, their surroundings become even more dire-looking. Wind knocks down a streetlamp, releasing a spray of sparks. The waters of the inlet rise, flooding the street. But Cantore and Bojorquez hold their ground. They're walled off from the flood by what seems to be a transparent force field. You can see the water get higher everywhere except for a circular zone around the two men. They continue their discussion about climate and weather dangers, noting how strong winds help blow seawater ashore in the phenomenon known as storm surge. "As sea levels rise, storm surges will move further inland," Cantore says.

The Weather Channel does not, of course, possess force-field technology. But it does have a different kind of magic: realistic-looking three-dimensional computer graphics that can insert meteorologists like Cantore into virtual scenes. The Weather Channel calls this approach immersive mixed reality, or IMR. It's a high-tech evolution of traditional green-screen technology, which has for decades been used to display weather maps or radar images behind meteorologists and is so pervasive that even TikTok offers a variation for anyone with a smartphone. IMR combines that ability to splice a person's image into footage with three-dimensional computer graphics that create virtual scenes. In one memorable example, the producers dramatized the destructive power of a large tornado by dropping a car next to Cantore, who obligingly cowered as the red sedan seemed to fall from above and strike the studio stage, sending virtual debris flying everywhere. That tornado segment helped the Weather Channel earn an Emmy for Outstanding Science, Medical, or Environmental Report.[169]

The technology makes for undeniably great television. But in the context of extreme weather dangers, IMR can educate as well as entertain. Many weather experts I talked with cited the Weather Channel's immersive videos as an important achievement in conveying weather risks to the public. The virtual-reality videos show hazards unfolding step by step, giving the on-air meteorologists the ability to pause the action and comment. It's particularly valuable when it comes to the threat of flooding, whether in the case of storm surge from a hurricane's ferocious winds or flash floods that occur when heavy rainfall soaks an area too quickly for the water to drain harmlessly away.

Once again, I'm reminded of that refrain when hurricane experts talk about safety: It isn't the wind that kills—it's the water. Meteorologists know that people consistently underestimate the dangers from flooding. According to the Centers for Disease Control, which tracks weather fatalities, the leading cause of flood deaths is driving a car or truck into water that turns out to be too deep for safe passage.

You can get a read on this yourself with a simple thought experiment. First consider how you would behave if your local forecast included a winter-storm watch and a prediction of possibly a foot of snow the next day. In most places, parents would prepare for their kids to be home from school, commuters would seek to work from home to stay off the roads, and many people would stock up on groceries. Next, think about hearing that a flash-flood watch has been declared. Would you consider canceling plans the next day or contemplate alternative routes for a car trip? Or even walk to the end of your driveway to clear leaves from the street's storm drain? The National Weather Service has been publicizing flood awareness since 2004 with a "Turn Around, Don't Drown" campaign. But communicating how quickly waters can rise to life-threatening levels remains a challenge.

"When you're in a storm like Sandy, you can't describe that to anyone. You see the video and it's nothing like it is in person," says Stephanie Abrams, a meteorologist at the Weather Channel and a familiar face to viewers from her years as cohost of its *America's Morning Headquarters* program. "It's like when you go on a trip and try to describe Mount Rushmore or Old

Faithful. It's so hard to describe what that feels like in person." Abrams tells me she and her colleagues have all reported from the field during these intense events, so they're well aware of how dangerous the floods can be. The IMR segments serve as a tool to underscore that for viewers. "It helps us explain to someone, 'This really is where the water will be.'"

To see the technology in action, I visited the Weather Channel's home in the suburbs northwest of downtown Atlanta. When I arrive a few minutes after six a.m., a staffer asks me to power down my cell phone lest an errant beep or buzz cause problems for the broadcast. We stop first at the control room for *America's Morning Headquarters*, which is already underway. A video wall at the front of the room shows more than a hundred different feeds, ranging from live cameras to computer maps and graphics. On one panel I spot Abrams, standing by for her next segment. Meanwhile, Cantore talks on the live broadcast feed. His voice booms out from the control-room speakers with an update on Hurricane Earl: "Intensifying Earl, eyeing up Bermuda," he says. "Will it cross over the island? Probably not." (Indeed, Earl steered well east of Bermuda.)

From the control room I make my way to the main studio, passing through a door with an ON AIR sign glowing above it. It's a cavernous space, housing multiple sets that the meteorologists can move among during the course of the show. Most of these look like familiar TV sets. There's a standing-height desk as well as a sofa-and-coffee-table setup with a talk-show feel where the broadcasters can sit for a conversation or interview a guest.

Over on the opposite side of the studio, though, the set looks very different. A neon-green backdrop fills every square foot of this space, extending high into the light-studded rafters and down to cover a sizable area of the studio floor. This is "Greenland," as crew members have nicknamed it. It provides the stage for immersive mixed-reality segments. This morning, Abrams stands amid the field of green, her bright pink dress contrasting sharply with the background. "We can transport you to Jacksonville, where you're going to start with the sun. You're going to want to be in the shade," she tells viewers as a camera mounted on a boom swoops down toward her. "Then we switch over, of course, to seeing the thunderstorms as we head into the afternoon, and some of these could actually be quite fierce."

In the studio, the scene presents a minimalist pink-and-green tableau. But on the monitor showing the broadcast feed—and for viewers at home—Abrams appears to be standing in Jacksonville, near the edge of the St. Johns River, with the Acosta Bridge behind her. As the segment begins, bright sun lights the river and buildings on the opposite bank, while shadows from trees gently sweep the ground, the outlines of individual leaves visible. The perspective creates an illusion of three dimensions, making it seem as if Abrams is standing outdoors at the location. Then, once she mentions the likely afternoon storms, dark clouds blossom behind her, followed in seconds by rain and lightning—all computer-generated.

The Weather Channel's virtual worlds owe a debt to computer gaming, which advanced 3D graphics, and to sports broadcasting, which ushered in some early examples of mixed reality, such as a yellow stripe superimposed over the football field to indicate the yard for a first down. "Sports did a lot of innovative industry firsts," says Michael Potts, senior vice president of design and visual innovation at the Weather Channel, a job that makes him the graphics guru. "There's a lot of money in sports, so it's at the bleeding edge." Potts should know: He came to the Weather Channel in 2013 after more than a decade in sports broadcasting.

"The core focus is storytelling and helping connect our audience with the stories we want to tell," Potts tells me. "The typical broadcast for so long was a person and a map, a very traditional way of conveying information. But how do we connect to the story behind the forecast? And how do we visualize that in a way that's meaningful and impactful to our audience? That's been my mission." Potts looks back at that early tornado segment that dropped a virtual car next to Cantore. "We're trying to tell simply that these storms are so powerful, they can do these things to everyday objects," he says. "You could never go out and film that."

The Weather Channel launched in 1982, the year after MTV's arrival heralded the ascendancy of targeted cable channels.[170] In 2022, the Weather Channel averaged about 160,000 viewers during prime time in the Nielsen ratings, according to *Variety*. That's well below the giants of general-interest news on cable but unequivocally the top spot among weather competitors such as AccuWeather and the newcomer Fox Weather, which launched in 2021. Some people tune in daily; others flip to the Weather Channel when there's a storm on the way. Whether the forecast calls for a major hurricane, a tornado-filled day, or a winter blast, Abrams, Cantore, and their colleagues can be counted on for deep, knowledgeable coverage—with some lighter fare mixed in. On the day of my visit, to mark the end of kids' summer vacations, *America's Morning Headquarters* included a look at elementary-school photos of the meteorologists.

Not all the Weather Channel's communications efforts have been universally welcomed. In 2012, it announced a plan to assign names to major winter storms, similar to the process for hurricanes. Weather Channel meteorologists argued that naming major winter storms would make it easier to communicate about them and would help increase public awareness when a snowy blast was on the way. But some experts, including officials at NOAA, criticized the plan, describing it as too subjective compared with tropical-storm naming, which gets triggered when a cyclone reaches a specific wind speed. Some didn't care much for a nongovernment entity jumping into the name game, and others complained that it was just a way for broadcasters to hype up weather. Still, the Weather Channel developed a set of criteria for the naming process, and many members of the weather community now believe winter-storm names add to effective communication, even if they'd rather see an official agency in charge.

There's no question the Weather Channel's reach into millions of US homes via cable providers makes it a key communicator of weather information. In 2022, it launched Weather Channel en Español, the first 24/7 streaming weather service in Spanish. Sussy Ruiz, vice president and editor in chief of the service, says the goal is to provide everything from routine weather forecasts and coverage of dangerous events to spotlights on climate change and environmental justice. Ruiz, a veteran of Univision and Telemundo, says that Hispanic populations are generally underserved and tend to be more vulnerable. "The sooner we can inform them about something happening and the resources available, the better chance of responding they will have," she says.

Of course, given the size of the audience—Spanish is the second most common language in the United States—there are viewers and profits at stake. But that doesn't diminish the value, particularly at a time when NOAA has begun to recognize some of the disparities. For instance, a study published in 2022 found that Spanish speakers in the United States were less likely than English speakers to understand the terminology of a tornado watch.[171] Among other recommendations, the authors said that rather than use *aviso*, which means "notice," for warnings, official messages should use *alerta*. "When translating emergency messages from English to Spanish, practitioners should consider the translation of the *meaning* rather than the *word* when developing consistent, effective, and inclusive multilingual communication," they wrote. In 2023, the National Weather Service moved to tap artificial intelligence to broaden the availability of translated information in different languages.

Public safety puts the onus for communicating weather threats on public officials. That's as it should be. But some of the work at the Weather Channel, from streaming in Spanish to creating lifelike scenes of surging water, shows how much room remains for innovation. "I think we're very good at forecasting the weather, so much better than we get credit for," Cantore tells me minutes after a hectic morning on the air. "But people, for whatever reason, don't always absorb the message. We have to go in there and lay out the potential impacts and lay out what that means to you." The answer, he says, is to keep working on improving how that information gets transmitted. "Is there a better way to communicate the message? Absolutely. Both visually and what we say with words? Absolutely."

The Friday-afternoon light begins to soften as the weekly meeting at the ECMWF winds down. The scientists and staff gathered in the Weather Room and connected over videoconferencing have heard Tim Hewson's forecast calling for a cold weekend in the Reading area. Any office worker would recognize the rustle of a meeting about to break up. But Hewson has a bonus item on the agenda: a forecast by Pangu-Weather.

At the mention of that name, the audience seems to perk up and click back into focus. To many people, the term wouldn't mean much. Those versed in Chinese mythology might recognize Pangu (or Pan Gu) as a legendary figure said to separate the earth from heaven or, more philosophically, yin from yang. But to a roomful of meteorologists, Pangu refers to an artificial intelligence system for predicting the weather developed by Huawei—the global technology giant based in China—and, by extension, to an enormous development looming over the future of meteorology.

Artificial intelligence and machine learning have already begun to disrupt multiple fields and industries. These technologies promise to perform all kinds of tasks but stir fears of economic disruption and a future filled with misinformation in the form of realistic fake images and videos. For the sciences, AI techniques raise complicated questions. They can help researchers find answers and speed up processes. But it can be difficult to ascertain how an AI has determined its results, a drawback that is in fundamental opposition to the goals of the sciences, which seek to understand the world. The venerable journal *Nature* surveyed more than sixteen hundred researchers on the topic in 2023, and respondents noted the potential for faster data processing and computations. But they also pointed to risks. The most frequently specified negative impact was "Leads to more reliance on pattern recognition *without understanding*" (emphasis mine); nearly 70 percent of those surveyed cited that.[172]

For meteorology, AI offers a host of tools and potentially a radical alternative to numerical weather prediction itself—a possibility that shakes the foundations of modern forecasting. So it's no wonder that Tim Hewson's quick meeting ender on Pangu grabs the audience's attention. He displays two sets of seven-day weather maps side by side, one from the ECMWF's high-resolution model and one produced by Pangu. They look, at least in this specific example, remarkably similar. But even though the two methods arrived at similar results, the resources needed to produce those results differed enormously. Running any of the big global NWP models at NOAA or the ECMWF or elsewhere typically requires minute after minute of computation on some of the world's most powerful and expensive supercomputers. The Pangu result, Hewson tells the audience, took less than one minute on a single-processor machine. To drastically simplify the difference, think of it this way: The AI approach can get a speedier result because it predicts what the future will look like based on patterns in the data about the current state of the atmosphere rather than by simulating what's happening step by step. In other words, the AI makes something of a guess—an incredibly sophisticated one—instead of performing all the heavy number crunching of NWP. But it can't show its work.

As people begin to drift out of the Weather Room at the meeting's end, a few make a beeline toward one scientist: Matthew Chantry, a young mathematics Ph.D. His official title is machine-learning coordinator. That means Chantry serves as the ECMWF's researchers' point person for AI.

(He has counterparts in the forecasting and computing groups too.) After seeing the Pangu-Weather forecast that Hewson popped on the big screen, some of his colleagues want to quiz him about it before heading off to their weekends. Chantry doesn't mind. He's tasked with figuring out how machine learning could—and, just as important, should—fit with the ECMWF's work.

"It's an extremely exciting time, and I feel very privileged to have this position right now," Chantry tells me. "We're starting to see some evidence that big machine-learning models might have a role to play in weather forecasting."

Indeed, 2023 showed hallmarks of a pivotal moment in machine learning for weather. Late in 2022, the researchers behind Pangu-Weather released an early preprint version of a paper describing their work and results.[173] The Pangu researchers went on to have their work published in *Nature* in July 2023.[174] In December 2022, a separate group of machine-learning researchers—this time at Google—delivered an early version of a paper describing success with their own system, GraphCast; their work appeared in *Science* in late 2023.

Over a span of just months, the two most widely respected scientific journals in the world reported on claims of dramatically successful results from machine-learning weather prediction. In *Nature*, the Pangu team claimed that its system "obtains stronger deterministic forecast results on reanalysis data in all tested variables when compared with the world's best NWP system, the operational integrated forecasting system of the European Centre for Medium-Range Weather Forecasts." In the article the Google team published in *Science* a few months later, they trumpeted GraphCast as "a key advance in accurate and efficient weather forecasting" and said the system had "greater weather forecasting skill" than the ECMWF's high-resolution model on ten-day forecasts.[175] No wonder people in Reading are so interested.

These declarations came amid a much broader tipping point in artificial intelligence. The year 2023 brought AI into the mainstream. Millions discovered the often uncanny conversational abilities of ChatGPT; others tapped AI image generators to conjure up professional-looking artwork from just a few words of description. News articles described amazing AI feats—and also relayed warnings from skeptics who argued that AI posed an existential threat to humanity. But doomsday warnings couldn't slow the speed of AI tools moving into mainstream use. Regardless of the apocalyptic predictions, rising concerns about AI-powered fakes fueling misinformation, and examples of ChatGPT making up facts in what techies refer to as "hallucinations," 2023 made clear that AI had arrived and had the potential to shake up practically every industry and field.

Of course, the stakes for AI weather forecasting are considerably higher than they are for AI composing your vacation-request email. Amy McGovern, a computer scientist at the University of Oklahoma, operates across both the computer science department and OU's School of Meteorology. She also leads an effort with a somewhat unwieldy name: the NSF AI Institute for Research on Trustworthy AI in Weather, Climate, and Coastal Oceanography. "Our center is focused on creating trustworthy AI," she tells me. "Which means that we're trying to dive in deep with our users and figure out what it means for them to trust the AI and what we need to do to adjust the AI and make it trustworthy."

The very nature of AI weather forecasts makes that trust a challenge. NWP models rely on well-established physical phenomena to simulate the atmosphere's behavior. But the machine-learning systems don't understand physics any more than ChatGPT has a sense of humor. In either case, it's a matter of pattern matching, whether looking at atmospheric data or completing a punch line. "The tension, I think, comes from the old-school modelers who feel that the dynamic models are physics and that physics can't be replaced by anything else," McGovern says. But she adds that it isn't quite that simple. Even advanced NWP models have their own kinds of shortcuts for activity in the atmosphere that's too small-scale for the model or contains unusual complexity; these get simplified in what's referred to as parameterization. Apart from the seeming broader inevitability of machine learning sweeping through society, McGovern believes the evolution of weather forecasting will require AI because of the need to deal with the exploding amounts of data from new sources such as microsatellite constellations.

The way forward for AI forecasting, McGovern says, will depend on exhaustive testing, a process that could take years and that will involve experienced forecasters evaluating the predictions as part of their real-world work, not simply comparing results in research papers. "It's a slow process," she says. "At some level, you want some slowness, because you already have a process that works. So before you go to something new, you want to make sure that it really, really works." She points to the procedures for drug approvals as an analogy. Getting regulatory approval for a new medicine can take years. "You want it to be proven," McGovern says.

Already, though, it's difficult to find any corner of the weather world that isn't at least experimenting with machine learning. NOAA has a center for artificial intelligence. IBM and NASA are collaborating on research for weather and climate applications. NVIDIA, the tech company that has evolved from a graphics-chip maker to a powerhouse in the booming market for processors that can efficiently run AI programs, has shown impressive results with its own weather system, FourCastNet. As I've described elsewhere, organizations and companies hope to apply machine learning to weather at nearly every timescale and at hyperlocal resolutions on targeted regions. But medium-range global forecasting continues to serve as the indispensable utility player in predicting the weather. And AI is already reshaping the field due to its relative speed and lower costs—attributes that may help make it more feasible for smaller nations with developing economies to up their forecasting game.

Before I left the ECMWF, I asked Florence Rabier, the director general, where she thinks AI technologies are headed. First of all, she notes what so many others have said: The acceleration of machine-learning progress makes predicting its eventual use a moving target. When the center released its most recent ten-year road map, in 2021, it mainly envisioned machine learning as something that could speed up specific parts of the NWP process. "We thought, well, we can just complement our model, but we'll never run the whole model that way," Rabier says. Then came word of the results from Pangu-Weather and others. "We saw all of these people from big tech running forecast models that actually looked pretty good. So we're revisiting our strategy. Maybe we can go further than we thought."

One vision of how machine learning fits into the world of global-forecast models runs much as Rabier alluded to regarding that 2021 ECMWF road map: Find components of the enormously intricate process of producing an NWP forecast that would benefit from getting sped up and where the results from machine learning appear solidly similar to the traditional computations, then plug in machine learning as a module. For an imperfect analogy, consider a talented chef preparing a gourmet meal, a process spanning everything from sourcing the right high-quality ingredients to gauging when a roasting chicken has reached the point of doneness. If a high-tech cooking robot came along, you might conclude that it could never reproduce the complex decisions a human makes about the freshness of herbs or the taste of a sauce—but it might handle chopping onions and carrots more quickly, at least as deftly, and with little risk to the outcome. Similarly, machine learning might effectively substitute for some of the methods of, for example, data assimilation while leaving calculations about atmospheric dynamics to traditional approaches.

But as the power of machine-learning forecasts becomes more clear, Rabier and other NWP experts see other possibilities. When it comes to producing forecast ensembles, conventional NWP and machine learning could work side by side, with a relatively small number of key ensemble members getting the full-on, supercomputer-exhausting, traditional approach at a high resolution. Other ensemble members—hundreds, perhaps, or more—could be generated by machine learning. Forecasters could look at the results to see if all those extra ensemble members suggest any weather possibilities that didn't show up in the NWP output.

This kind of arrangement makes particular sense when you think about computing as a finite resource. Every time the ECMWF or NOAA upgrades the supercomputers they use for the global-forecast models, they effectively increase their computation budget. The newest machines can perform more calculations in the same amount of time, but there's still a limit on how much they can calculate in any given period. So key decisions about numerical modeling involve deciding how best to "spend" that budget. You might devote more computer time to increasing the resolution of the model. You could decide instead to use a coarser resolution, spending the savings in computing power to run more ensemble members. In this sense, machine-learning techniques are much more resource-efficient. They produce a forecast much faster and with fewer computing resources.

To Rabier, that efficiency is significant in another way: energy savings and the environmental benefits that come with it. "I believe machine learning is going to play a very important role in the future to accelerate what we are doing and avoid us spending huge amounts of electricity and damaging the planet to run our computations," she says.[176]

But even if AI winds up enhancing or complementing NWP, Rabier doesn't see it replacing it. For all the potential of machine learning, numerical models remain the gold standard for weather forecasting because they simulate what actually happens in the atmosphere. Each new discovery in physics and each new source of data can bring those models closer to replicating what happens in the real world. Machine-learning systems are only as good as the data they've been trained on. That means they're hampered in their ability to identify a pattern that never occurred in that training data—no small issue when it comes to unusual events, particularly as climate change morphs weather patterns.

There's also the question of ensuring access to high-quality training data. Detailed historical information about the atmosphere comes from something known as reanalysis. For these vast repositories of past measurements to be useful, the data must be massaged into a consistent form. The ECMWF describes a reanalysis as "a blend of observations with past short-range weather forecasts rerun with modern weather forecasting models [. . .] globally complete and consistent in time." In other words, as the ECMWF puts it, "maps without gaps." It's a technical process, but one that's hugely important to the world of meteorology. The center maintains a reanalysis called ERA5 as part of its work for the European Union–funded Copernicus Climate Change Service. Not surprisingly, the quality of the ECMWF's model makes that ERA5 data especially useful. Both Pangu and GraphCast trained on ERA5. Rabier isn't shy about highlighting the debt these systems owe to the ECMWF. "The success of the machine-learning models all rely on our reanalysis," she says. "You still need the ground truth at some point."

Several months after I visited, ECMWF scientists launched their own experimental machine-learning forecast model. They named the newcomer the AIFS, for Artificial Intelligence/Integrated Forecasting System, a clear nod to the center's storied NWP model, the Integrated Forecasting System, or IFS. "Physics-based numerical weather prediction models, for us the IFS, are still key in all of this," the team wrote in their announcement.

7 Seasonal Forecasting

EARLY WARNINGS FOR DROUGHTS, FLOODS, AND FAMINE

The climate of Zimbabwe shifts as you move across the country. The nation is landlocked, with Mozambique situated between Zimbabwe's eastern border and the Indian Ocean. To the north and east, warm and wet conditions tend to dominate. But move to the south and west, toward Zimbabwe's borders with Botswana and South Africa and the Kalahari Desert, and the environment turns dry—or semi-arid and hot, according to its category in the Köppen-Geiger scale, a widely used system of climate classification.

Though Zimbabwe's economy benefits from a lucrative mining industry, agriculture dominates day-to-day life for many of its people. The twenty-first century witnessed a dramatic upending of the country's farming sector, which had historically been a powerful economic engine but, with the vast majority of productive lands in the hands of white owners, had a problematic legacy of colonialism. Reform efforts struggled. Then, in 2000, followers of revolutionary leader Robert Mugabe marched on farmlands, seized them by force, and ejected the white owners, sometimes violently. In the aftermath, land was redistributed to Black Zimbabweans who could then work the land as smallholder farmers. But the process was riddled with corruption. The upheaval ultimately devastated the productivity of the agriculture sector, taking a country previously considered a breadbasket for the region and blowing up a key driver of its economy. The result: hyperinflation and food shortages. It has taken years for agricultural productivity to recover.

Today, most farmers work relatively small plots in what's referred to as subsistence farming—the goal is to feed the family; they sell any surplus harvest. In those dry regions, places like the Gwanda District, where mining operations for lithium mix with agriculture, farming always presents a challenge. But in periods when the already meager rains fail to materialize, the effect can be disastrous. For example, when a lack of rainfall in 2015 kicked off a major drought, poor maize crops created a food emergency, with nearly four million people in the country in need of help, according to a Reuters report.[177] Kids were at particular risk. UNICEF warned that "nearly 33,000 children are in urgent need of treatment for severe acute malnutrition."[178] *The Guardian* reported schoolkids fainting from hunger in classrooms. Humanitarian agencies mobilized to provide food, water, and other assistance. The United States upped its aid contribution throughout the region, sending an extra $54.5 million for Zimbabwe, according to the US Agency for International Development.[179]

Such dry periods are a recurring hazard. In 2023, humanitarian groups once again raced to offset the dangers of drought-driven hunger and the follow-on consequences of a food shortage.

But this time, there was a major difference: The drought hadn't happened yet. Instead of waiting for the flagging harvests and reacting to the resulting shortages, a network of officials and organizations—ranging from agencies of the Zimbabwe government to international NGOs like the World Food Programme (WFP) of the United Nations—sprang into action. The impetus for this proactive surge was a long-range weather forecast.

Months ahead of the crucial growing season that begins in October and November and continues through January and February, officials scrutinized long-lead weather outlooks. When these forecasts indicated that a drought would probably materialize, the officials activated a previously agreed-on plan, drawing on humanitarian funds designated for such a scenario. They began communicating the risk to farmers through radio broadcasts. They tasked agriculture agents with spreading the word through communities. And they launched the distribution of seeds for drought-resistant crops at no cost to farmers, encouraging them to make planting decisions that would protect them in case of an arid season.

"When the seasonal forecast is out, we see how severe the drought, or the wetness, will be, and then we advise accordingly," says John Mupuro, a forecaster at Zimbabwe's national weather agency, the Meteorological Services Department, in Harare, the nation's capital. In August and September of 2023, Mupuro tells me, he and his colleagues worked with other organizations to make sure farmers were armed with the best weather information. "They would then go into groups to actually organize and talk about what they would want to do in the season, given the forecast."

It's called anticipatory action. And it's a simple concept: Aid money spent in advance of a crisis to help people prepare and avoid the worst effects can be far more potent than funds deployed once disaster has struck. That might sound like common sense, but only in the past decade have major aid agencies become willing to put their money behind it. There is, after all, some risk. If the forecast conditions don't actually occur, money and supplies could wind up deployed unnecessarily to one region while a problem brews elsewhere.

But weather scientists have demonstrated that long-range forecasts can provide valuable early warnings of unusual conditions weeks or months in advance—if they're properly interpreted and used. Figuring out how to put them to work promises new benefits across a number of important applications and industries. It also requires fluency in a key area of weather literacy: the probabilistic forecast.

In everyday life, most people want a yes-or-no answer to their questions about weather. Will it rain tomorrow? Will schools get a snow day on Friday? Will the weekend be too hot to play pickleball or work in the garden? More seriously: Do I live in the path of the hurricane predicted to arrive next week? If you've made it this far in this book, hopefully you have a more sophisticated understanding of what a good weather forecast can offer. Any meteorologist will tell you that forecasts represent probabilities, not certainties. Consumers of those forecasts would be wise to treat them as such. But it's human nature to want that yes-or-no answer, to hear that it will rain or it won't, to be told that the high will be 85 degrees. Meteorologists refer to these as deterministic forecasts. They contrast with probabilistic forecasts, which assign chances to different outcomes: On Friday there's a 20 percent chance of no rain, a 50 percent chance of less than half an inch of rain, and a 30 percent chance of more than half an inch. Still, even if the probabilities undergird every forecast, most people pay attention to the deterministic summary. Icon of the sun on your TV or smartphone screen? Nice day tomorrow! Cartoon lightning bolt? Get ready for rain.

For the most part, short-term deterministic forecasts work, at least for casual consumers of the information. All the advances in computer modeling, satellites, and radar as well as our understanding of the underlying physics of weather have extended our ability to predict. For instance, NOAA data shows that five-day forecasts of high temperatures are now off by just under four degrees on average—about as accurate as a three-day forecast from twenty years ago. When you push the time horizon farther out, the accuracy falls. Yet models keep improving, and machine-learning approaches promise to aid the pursuit of better forecasts in the ten-day range.

But on the scales of weeks and months, things work differently. Meteorologists refer to these predictions as subseasonal (for forecasts covering periods from two weeks to three months out) and seasonal (three months to two years). Weather experts sometimes use the shorthand S2S, for "subseasonal to seasonal," as an umbrella for these long-range outlooks. They can't offer the kind of pinpoint predictions we're used to from shorter-range forecasts. But—as projects like the drought effort in Zimbabwe show—they can be very useful if you understand what they can and can't do. Literacy in the language and concepts of probabilities is crucial.

Working with probabilistic forecasts amounts to making bets on the weather. As with any bet, the odds matter. An 80 percent chance of something happening feels close to a sure thing (even though, actually, it isn't). A 20 percent chance of something happening feels like long odds—but depending on what's at stake, you might still decide to cover your bases.

At NOAA, a unit called the Climate Prediction Center (CPC) handles

S2S forecasts for the United States. Examine any CPC forecast product and you'll see the probabilistic approach at work. For instance, the CPC produces temperature and precipitation forecasts for the period three to four weeks ahead. The Weeks 3–4 Temperature Outlook shows a map of the United States with broad, curving zones superimposed over it. The zones are color-coded in reds and blues to indicate whether the CPC expects temperatures in that part of the country to be above average or below; the darker the shade, the higher the probability.[180] Looking at a map issued in February for mid-March 2024, I can see that the band covering Wisconsin, Michigan, Pennsylvania, and most of the Northeast is a deep rust/maroon, which the map key labels as an 80 to 90 percent chance of above-normal temperatures. The same map colors Southern California and Arizona in a mid-blue that indicates a 55 to 60 percent chance of lower-than-average temperatures. A separate three- to four-week precipitation outlook divides the country up into areas where rain or snowfall are expected to be above or below normal.

The CPC publishes similar temperature and precipitation forecasts looking one month out, three months out, and even a year ahead. For the average person hoping for intelligence on conditions in the coming weeks, these forecasts aren't exactly user-friendly. For one thing, even guidance that temperatures have an 80 percent chance of being higher than normal doesn't tell you much if you aren't familiar with what's normal in the region you care about. (If you visit the CPC website, you can click through to tables of historical averages for different parts of the country.)

None of this is likely to help you decide whether to hold your big family reunion indoors or outdoors. Yet in any industry or application where small deviations from the norm can add up, these kinds of forecasts—making a bet on the weather—can pay off. Energy companies, for instance, can decide whether to lock in extra supplies of heating oil if the winter ahead appears likely to be colder than normal. The agencies that manage water reservoirs and systems can plan for shortages if the forecast calls for lower-than-normal precipitation. From retailers projecting demand for seasonal goods to contractors estimating when conditions could interfere with construction schedules, playing the odds on seasonal forecasts can be good business. But perhaps no one has cared more or sought answers longer than farmers.

In ancient civilizations that developed sophisticated astronomical observations, the desire to understand the timing of the seasons helped drive the invention of calendars. Thousands of years ago, Mayans and Aztecs are believed to have created calendars that would aid them in knowing when the time for planting maize crops had arrived. In the Han Dynasty, Chinese astronomers watched the skies and refined calendars to predict the arrival of key periods for agriculture.

In 1792, a Yankee schoolteacher named Robert B. Thomas published the first edition of *The Old Farmer's Almanac*, a guide that continues to be issued annually to this day. The publisher of the *Almanac* says it relies on a "secret weather formula" that can be traced back to Thomas's own methods: "Thomas believed the Earth's weather was influenced by sunspots [. . .] and this factored heavily in his forecasts."[181] In fact, sunspots can indicate where the sun is in a roughly eleven-year cycle between its maximum and minimum energy output, and climate scientists believe that can have an effect—though small—on earthly weather.

Today, subseasonal and seasonal forecasts tap a much deeper understanding of the atmosphere. But the annual rhythms that the ancients relied on still serve as a foundation, enhanced by modern observations. One relatively simple approach to forecasting the weather weeks or months in advance is called climatology. In climatology, you take years of weather data and calculate the average. If it's July and you want to know the likely weather in Dallas down the road in January, you can find that over the years, the typical high temperature is roughly 57 degrees and the low comes in at about 36.

Think of climatology as the weather world's game of craps: When you're rolling a pair of dice, over time the most frequent result will be a seven, but of course, on any given roll you might get anything from a two to a twelve. The variations in weather aren't usually quite so broad, so those historical averages provide valuable insights. So when meteorologists seek to create a useful subseasonal or seasonal forecast, they're looking to beat the house by out-predicting climatology.

One way to do that is to analyze big recurring patterns. In the vast envelope of air that is the Earth's atmosphere, there are disturbances both small and large. The small ones move quickly: A thunderstorm that dumps rain on an area the size of Manhattan can form and then dissipate in fifteen minutes. Larger phenomena proceed more slowly: A hurricane three hundred miles wide evolves over days. But the truly enormous patterns, which can span an ocean or a continent, unfold gradually over weeks or months. They can be so powerful that they alter the weather on the opposite side of the planet. Meteorologists and climate scientists refer to these globe-crossing links as teleconnections. El Niño may be the most well known, though there are others. And studying them helps makes subseasonal and seasonal predictions possible.

To understand how these big patterns work, it helps to remember some of the fundamental forces acting on the atmosphere. One is the sun, which warms the air, but not uniformly. Thanks to the spherical shape of our planet, sunlight heats the equator more than the poles, where light hits the atmosphere at a more oblique angle. Another key force is the Earth's rotation, which is responsible for the Coriolis effect. If you examine a point on the planet's surface near the equator, you'll find that it moves at a faster speed than a point near either of the poles, since the distance around the Earth is much greater near the equator. Because of the Coriolis effect, air near the equator appears to deflect to the right as it moves northward toward the Arctic. (If you want to visualize this better, the UK Met Office has a helpful series of videos on YouTube.)[182] The upshot of these factors—the unequal heating of the Earth and the impact of rotation and the Coriolis effect—is the large-scale circulation of the atmosphere, including the jet streams and the permanent prevailing winds referred to as the trade winds.

A University of Chicago meteorologist named Dave Fultz helped deepen understanding of these globe-spanning effects with a series of investigations in the 1950s known as dishpan experiments. To reproduce basic conditions of the spinning Earth, Fultz used a cylindrical container of water—the dishpan—and set it rotating on a turntable. The water represents air, since both display fluid behavior. Heat was applied at the edge of the dishpan to mimic the greater effect of the sun at the equator, and the center of the pan was cooled as a stand-in for a chilly pole. When dye or metal powder was dropped in, the currents in the water made it circulate and patterns emerged. These experiments showed how the jet streams shape the weather as they undulate between the equator and the poles.[183]

Add one more key process and you have the basics of global circulation: air-sea interaction. Around the world, the atmosphere and the oceans mingle with each other constantly. The power represented by this dynamic becomes obvious when you view an image of Earth as seen from space. Water covers an estimated 71 percent of the planet's surface, with oceans representing all but a fraction of that amount. Wherever air touches water, tiny transactions take place. Oceans accumulate thermal energy from the sun's rays. Warm ocean water gives up some of its heat to the atmosphere. Seawater evaporates, pumping moisture into the air. Changes in air temperature alter the pressure of the atmosphere's gases. Combine those ingredients—changes in the atmosphere's moisture, temperature, and pressure—and you have the underlying drivers of the planet's weather.

Comprehending the air-sea connections involves something of a chicken-or-the-egg viewpoint. Does the ocean change the atmosphere or vice versa? The answer, as you might expect, is both. For instance, consider the power of wind to influence the ocean. Air skimming over the surface moves the water, creating waves. On a small scale, those waves might toss a boat around. But over a large area, winds drive a much more expansive phenomenon called upwelling. As winds push water to the side, they make room for water below the surface to rise. The deeper waters are colder, which means that upwelling changes the sea's surface temperature—one of those critical variables in the air-sea interactions. (Of interest to fish and those who want to catch them, the colder water also lifts nutrients from below, feeding into the bottom of a food chain that ultimately sustains everything from flounder and shrimp to fish-hunting birds.)

Step back further and you can observe wind and waves driving persistent ocean-wide patterns known as gyres. Here too the Coriolis effect makes itself known. In the Northern Hemisphere, the major gyres display a clockwise pattern when viewed from above; counterclockwise movement dominates in the Southern Hemisphere. In the North Atlantic, for instance, water on the west side of the ocean flows toward the north in the Gulf Stream, sending warm Gulf of Mexico waters up the Eastern Seaboard before they swing east, or clockwise, toward Europe. On the European side of the Atlantic, currents turn south toward Africa before eventually curling westward toward the Gulf of Mexico, completing the loop.

No matter the weather outside your window, these large patterns—persistent ocean currents and big atmospheric circulations—are always in play, with air and water exchanging heat, moisture, and energy. Which brings us back to El Niño.

The term *El Niño* is believed to have originated with Peruvian fishermen in the nineteenth century or earlier. Under normal conditions, they enjoyed bountiful catches thanks to currents and upwelling that brought cold water packed with nutrients to the coast of South America. But sometimes, the fishers noticed, the ocean would change. Warm water would move in, depleting the stocks of anchovies. Because the warm water typically arrived near the end of the calendar year, they associated it with Christmas and described it as El Niño—"the boy"—after the Christ child. Usually after some weeks, the cold waters returned and the fishing nets once again filled.[184]

Sometimes, though, the warm pattern persisted for months. This extended disruption is what most people mean today when they refer to El Niño: an unusual shift in weather patterns marked in part by warmer-than-normal waters off the western coast of South America. A peculiar shift in the opposite direction—where those same waters turn strangely cool—is known as La Niña ("the girl"). And meteorologists now understand that both are part of one of the most powerful drivers of global weather: the El Niño Southern Oscillation, or ENSO. (Weather scientists pronounce it *"ehn-*so.") ENSO refers to the overall pattern that moves between the El Niño effect, the La Niña effect, and a neutral stage.

The ENSO dynamic ties together those ocean and atmospheric circulations. Normally, trade winds over the Pacific near the equator blow toward the west, allowing that upwelling of cold water off South America. But occasionally, changes in pressure diminish the strength of those trade winds or even reverse them. This back-and-forth of air pressure is the southern oscillation part of ENSO. When that shift happens, the prevailing equatorial winds blow toward the east, moving warm Pacific waters toward South America and bringing the extended warming noticed by those Peruvian fishermen in an El Niño event.

What happens in Peru doesn't stay in Peru. An ENSO event, whether El Niño or La Niña, alters weather worldwide and lasts for months or years. Each stage brings a set of characteristic effects, which is why seasonal forecasters watch so closely for an emerging ENSO event. In the El Niño phase, low pressure over the northern Pacific pushes the jet stream—that powerful river of high-altitude wind—toward the south as it passes over the eastern Pacific. In the United States during winter months, the dipped jet stream carries moist air toward California and onward across the southern states, while the Northwest experiences warmer temperatures. Not every El Niño produces strong effects, and in many places they don't have much impact on the weather. But around the globe, other regions typically see a deviation from average conditions. A strong El Niño can bring unusually wet weather to equatorial East Africa from October through December. And in southern Africa, El Niño can turn the weather dry from November through March, raising the risk of drought and all its consequences for farming and access to water in places like Zimbabwe and Mozambique.

For all of El Niño's notoriety, the ENSO isn't the only large long-term phenomenon changing the weather. In the Pacific Decadal Oscillation (PDO), scientists have observed a pattern that can persist for ten years or more (hence *decadal*) before reversing. Attributed in part to changes in a persistent area of low atmospheric pressure near the Aleutian Islands, the PDO exhibits a warm phase in which warm water moves toward the eastern Pacific and the west coast of North America. In the cool phase, the PDO sees cold water in the eastern Pacific and warmer water on the west side of the ocean. Depending on which phase the cycle is in, it can lower or boost temperatures and precipitation in specific regions—including the United States, where the PDO can shape how cold and wet winters are in different parts of the country.

Another major phenomenon is the North Atlantic Oscillation (NAO), which can flip back and forth in periods lasting anywhere from days to months, altering the weather in the eastern United States as well as western and central Europe. The Arctic Oscillation (AO) is similarly brief in its phases, particularly when compared to the ENSO. The Indian Ocean Dipole (IOD) can dump greater than average precipitation on eastern Africa and dry up the weather over Australia. Adding to the complexity of all these big patterns, they can also interact with each other. The ENSO, for instance, affects the air pressure over the Aleutians, which can flip the PDO.

Then there's the Madden-Julian Oscillation (MJO). Even though it displays cyclical behavior like the ENSO and the NAO, it exhibits an additional characteristic that makes it more complicated: It moves around rather than staying in one place.[185] It circles the globe, tracing a path within the tropical latitudes and moving from west to east at a pace that takes about thirty to sixty days to go around the world. The MJO has two components; one is a zone where rainfall is higher than normal and the other where it's less than normal. Depending on the location of these zones at different times of the year, the MJO can boost hurricane activity or chill temperatures during winter in parts of the United States. The MJO can also affect the ENSO as it wends its way around the Earth.

Put all of these together—persistent ocean currents, large and relatively slow atmospheric patterns, and the constant interplay between the oceans and the air above them—and you get a complex picture but one that nonetheless displays some long-lasting and slow-moving systems that can help forecasters predict the weather weeks ahead of time. Understanding them better and gathering more observations that indicate the current state of each cycle gives meteorologists valuable tools to produce subseasonal and seasonal forecasts. The bigger challenge may be helping people to put those forecasts to work.

"The subseasonal-to-seasonal (S2S) predictive time scale [. . .] is at the frontier of forecasting science." So declared a group of sixty scientists from around the world in a 2022 paper in the *Bulletin of the American Meteorological Society* (*BAMS*).[186] "There is, however, a 'knowledge–value' gap, where a lack of evidence and awareness of the potential socioeconomic benefits of S2S forecasts limits their wider uptake." This group of experts, from national and international forecasting agencies to researchers at Columbia University's influential International Research Institute for Climate and Society to big corporations, came together to present a collection of case studies that demonstrated the promise of longer-range forecasts if the right people and groups could figure out how to use them.

Preventing malaria, for instance, emerged as one of the ways for S2S to bolster public health efforts. Because the mosquitoes that carry the malaria parasite are sensitive to temperature and moisture, warm and humid weather can promote the spread of the illness, which the World Health Organization estimated killed more than 600,000 people worldwide in 2022. Researchers worked with Nigeria's national meteorological agency, which operates a real-time weather-monitoring system to gauge the likelihood of outbreaks, essentially using current weather conditions to forecast disease potential in the weeks ahead. The researchers analyzed the use of S2S predictions in the outbreak forecasts and concluded they would add value in "supporting early identification of malaria hyperendemic areas, as well as prompt mobilization and intervention by the responsible health department, at least a month before the outbreak of the disease."

Another application is renewable energy. Weather obviously affects demand for energy. When it's hot, people will run their air conditioners; when it's cold, people need power for heat. But as economies around the world attempt to transition to renewable energy sources that lower carbon emissions, weather represents a critical factor in the supply equation too. Cloudy weather means less available power from solar; breezy conditions give a boost to wind farms. When a European research group worked with energy-industry companies to devise a tool that incorporated ECMWF subseasonal weather forecasts, they found a variety of potential benefits. For instance, utilities could plan to take power plants offline for needed maintenance when demand was projected to be low. Advance word on possible snowfall that could cover solar panels would give operators a chance to prepare for clearing them or to tap other sources.

The *BAMS* paper outlined other uses for subseasonal and seasonal forecasts, ranging from guidance on how to manage water reservoirs to assisting with preparations that could avoid unnecessary deaths during severe cold snaps. The authors emphasized that making the most of S2S forecasting involves putting meteorologists together with those who can provide domain expertise in whatever field is at issue. The work needs to be integrated, rather than having the weather scientists simply deliver a forecast.

The paper also acknowledged obstacles. Perhaps most prominent is the challenge of getting end users and decision-makers weather-literate enough to be comfortable with the probabilistic nature of long-range forecasts. "The S2S forecasting time scale is therefore a new concept for many users," the researchers stated, adding that "incorporating probabilistic ensemble S2S forecasts into existing operations is not trivial. S2S forecasts do not produce a 'go–no go' answer of what a user should do." Ultimately, putting S2S to effective use requires a longer view and an acceptance that any single forecast may not come to pass. But if they're right more often than not, if they provide a better look forward than simple climatology, then making bets on them should pay off over time.

To learn more about where subseasonal and seasonal forecasting are headed, I talked with Tim Stockdale, the long-range forecasting guru at the European Centre for Medium-Range Weather Forecasts. Stockdale is the first to admit he has an interesting role at the ECMWF, because *medium-range* is right there in the name of the institution. In keeping with the ECMWF philosophy of a single numerical weather-prediction model, seasonal forecasts there are built on top of the IFS, the cornerstone numerical weather-prediction model of the center's work. "We have one forecast system, and we use that same forecast system for all timescales," Stockdale reminds me.

Different factors matter more at different forecast ranges, Stockdale explains. For the short- and medium-range forecasts, the quality of the initial conditions—that snapshot of the environment that the model works from to look progressively deeper into the future—matters a great deal. But moving forward in time, Stockdale says, the actual model holds more sway. The longer and slower processes in the atmosphere drive the future conditions. "And so, in a way I'm the most demanding person in the building in terms of the quality of the model, because errors that other people think are small, actually over a long period of time, they can really cause damage," he says. "The biggest single thing that will improve over time our forecasting on the timescale of weeks to months is the quality of the model."

For instance, moisture in soil moves into the air through evaporation, affecting temperature and humidity, with significance for longer-term weather patterns. But the payoff for representing variations in that process may not be worth the computing cost of the extra calculations in a short-term forecast. Stockdale's recent research has focused on another component: aerosols, the tiny particles of matter or liquid in the atmosphere. Stockdale says these are important for two reasons: They can absorb or scatter the light from the sun, and they interact with clouds. Modeling the behavior and effect of the aerosols in a complicated way may not be worth the expense of the additional computations in the short term but could make longer-range forecasts more accurate.

How far out does Stockdale believe we can usefully forecast the weather? More than you might think—as long as those forecasts are used with the proper appreciation for their benefits and flaws. "You can maybe go into a two-year timescale," Stockdale tells me. But some periods will be more predictable than others. Whether some of those big patterns like ENSO and the Pacific Decadal Oscillation are in a stable or transitional phase, for instance, will affect predictability. Weather scientists gain insights into things like this by running their latest models on past conditions and comparing the results to what really happened. "Over the twentieth century, there are some decades where predictability is higher or lower," Stockdale says.

Figuring out whether you're in a more predictable phase can help you decide how much to rely on the seasonal forecast. An important tool for doing so is the ensemble forecast, in which the numerical model runs multiple times with slightly different factors in each run. If the ensemble members vary wildly, that's a caution sign. If they group together, that's encouraging. "When we run our ensemble forecasts for El Niño, there are times where it's quite tightly constrained," Stockdale says. "So the predictability of something like El Niño, we understand it relatively well."

Even as scientists refine their understanding of the atmosphere and its patterns, big gains are already possible from smartly deploying S2S predictions. "It's one thing to try to improve the forecasts," Stockdale says. "That's what I'm most passionate about. But the application and use of those forecasts, that's something that really has a lot of scope for further development." Predicting the likelihood of droughts or floods in Africa—exactly the point of forecast-driven anticipatory-action programs—is a key example for Stockdale. Meteorologists will keep developing more accurate forecasts, but putting subseasonal and seasonal predictions to work can't wait. Says Stockdale: "With seasonal prediction, the question of how it translates from the forecast to the use of it, and the impact on society—that's quite a big thing."

For the farmers in Zimbabwe who depend on their crops to feed their families, support livestock, and ultimately earn some income, a key link in the system to help them put forecasts to work is Moffat Ncube. He is an expert with the government's Department of Agricultural, Technical and Extension Services, better known as Agritex. Ncube trains extension officers who work with farmers across the country and spends time in the field himself, listening to farmers' questions, hearing about their challenges, and sharing knowledge that can help. Ncube explains to me the calendar that dictates the planting season for these farmers and where the forecast fits in.

"Before the season, that's when the Meteorological Services Department gives us a forecast of how the rainfall is going to be," Ncube says. "They may want to reduce maize. Maize is not that drought-tolerant." Instead, he says, they can consider crops such as cowpeas, the legume known in the United States as black-eyed peas. The cowpea provides plenty of protein and other nutrients while thriving with relatively less water. "So now the farmers have been told and they are able to say, 'If we are going to be having lower rainfall than normal, then we need to change how we plant,'" Ncube tells me.

Ncube emphasizes the agency of the individual farmers in deciding how to respond to the forecast conditions. This is considered one of the benefits of the anticipatory-action approach. Aid is, of course, an important part of the equation, and the program for Zimbabwe includes funds to purchase and distribute cowpea seeds and fertilizer for affected farmers.[187] But by sharing weather predictions along with agricultural insights, officials bring the people most affected into the decision loop—a more sustainable and more respectful alternative to leaving them to wait for aid after droughts and food shortages. "It's being equipped, being empowered," Ncube says. "Their facilitation is such that at the end of the day, the farmer is able to choose the enterprises that are most likely to succeed."

Another link in this chain is the media. One of the partners in the anticipatory-action program is the Campus Radio service of Great Zimbabwe University. Like campus radio stations everywhere, the GZU service puts students on the air. But it also plays a public-service role, drawing on university experts to share knowledge and "help the community to solve existential problems and to proffer solutions through research and innovations," as its mission statement describes it. For GZU Campus Radio, helping to disseminate news related to the anticipatory-action program dovetails with a broader goal of bringing critical information and context about climate change to the smallholder farmers affected by the environment.

Golden Maunganidze, the station director, says these information efforts must work to overcome the tendency of some individual farmers to hew to traditional wisdom and guidance passed down from earlier generations about when and what to plant. "Most of them, they've not been relying on official information channels, from research institutions, government, stakeholders, and other interested parties," he says. "And yet, these are people who are being affected by serious changes, in terms of climate, and also in terms of their general day-to-day interaction with the environment."

Part of that work relates to the same challenge that everyone faces with long-range forecasts: helping people wrap their heads around probabilities and uncertainties. Champion Mudavanhu, one of the producers for GZU Campus Radio, puts it this way: "No forecast is cast in stone."

In July 2023, with scientists at the Meteorological Services Department cautioning that the El Niño conditions would likely translate to a drought, the radio station spread the word—but to an audience containing skeptics of how accurate the weather forecasts could be. When October brought some rains—an occurrence that might ordinarily herald the approach of the growing season—some farmers assumed that conditions would be normal. "But the Met Services Department came in and clarified that those were *not* the rains that indicated the start of the season," Mudavanhu tells me.

This problem is familiar to meteorologists everywhere; they all face dismissive comments whenever a forecast fails—or even just appears to fail—to pan out. Everyone in the chain, from the Agritex extension officers to the people speaking over GZU Campus Radio, had been working to communicate that the drought wasn't a certainty but a likelihood that demanded smart preparation. "The idea now was that people generally distrusted information from the Met Services. At some point they would say, 'You guys say this, and then, you know, it's not happening,'" Mudavanhu says. "So we had to really explain to people how the Met Services work with the data and how all this comes together to be useful in terms of farming or in terms of how people live through their day-to-day lives."

Anticipatory-action programs may be relatively new, but already there's evidence supporting their value. A 2018 study for the US Agency for International Development that looked at drought problems in Kenya concluded that early action would save hundreds of millions of dollars on the cost of aid itself and even more when economic knock-ons like livestock losses were included. "When these estimates are applied to total U.S. Government (USG) spending on emergency food aid in Kenya, the USG could have saved US$259 million over 15 years in direct cost savings, or 26% of the total cost of emergency aid," the study found.[188] "Incorporating the avoided losses to households, the model estimates net savings of US$1.2 billion."

In Nepal, the World Food Programme analyzed a forecast-based response to flooding versus traditional disaster response in a 2019 paper. The study found that the average cost of a typical flood emergency—one that involves 175,000 people—comes to $31.5 million when addressed through a traditional response but only $10.1 million if forecasts and anticipatory actions are deployed. As the report stated: "Growing evidence suggests that robust early warning systems based on credible scientific weather forecasts, together with preparedness and anticipatory action protocols, can add significant value in increasing community resilience and reducing the need for humanitarian assistance."[189]

The stakes, of course, are greater than items in a budget. The aggregate numbers don't do justice to the human suffering that occurs when floods displace a community and wipe out crops or when a drought leads to a famine. When the United Nations Office for the Coordination of Humanitarian Affairs (OCHA)—which works to coordinate efforts that include WFP responses to food insecurity—published a summary of the anticipatory-action plan in 2023 for the expected El Niño effects in Zimbabwe, it noted some of the consequences from the 2015–2016 El Niño drought. Hunger and malnutrition were only part of the crisis. People turned to "unsafe water sources for drinking and domestic use, even sharing these sources with livestock," according to the report.[190] It continued: "These shortages had extensive adverse effects on health, nutrition, school attendance, clinic operations, and the risk of violence." Because women and girls are typically the ones to retrieve water for a household, they "faced elevated risks of gender-based violence (GBV) as they had to undertake longer journeys in search of water."

Anna Lena Huhn has been working on anticipatory-action efforts since their early days in the mid-2010s. Now she coordinates anticipatory-action and early warning efforts for southern Africa at the UN's World Food Programme. She says the forecast element has required a lot of education and persuasion. "The first years there was zero interest from the wider [humanitarian] system and skepticism from donors," she says. "Why should we invest? Why should we spend money when we don't know what will happen?" Over time, the doubts have faded, she says. Naysayers have come to accept that, even though any individual forecast might fail to prove accurate, the cumulative benefits support the approach. "It's been shown that even if we activate in vain—even if we disperse funding, implement actions reaching vulnerable households—that it would take up to six false alarms to reach the cost of one conventional late emergency response," Huhn says.

Future progress will come with further popularizing the approach and deepening the participatory process by which forecast triggers and the related actions are agreed on. But in many places, it also means helping countries invest in their own meteorological infrastructure. Directly to the east of Zimbabwe, the nation of Mozambique includes regions vulnerable to low rainfall in an El Niño event. The WFP provides support through anticipatory-action programs there too. It has also worked over several years with the Mozambique National Institute of Meteorology to digitize historical weather data, install weather stations, and support seasonal forecasting. "These are investments in the weather-observation and climate-observation network," Huhn says.

Effective anticipation also requires combining the meteorology with other forms of forecasting and modeling. "The more recent move is to be thinking more about impact-based forecasts," says Liz Stephens, a professor of climate risks and resilience at the University of Reading, which has close ties to the nearby ECMWF. "Not just saying, 'Okay, we're going to get one-twenty-kilometer-an-hour winds,' but thinking: 'Who is the population that's exposed? What conditions are they living in?'" That requires factoring in not just meteorology but also hydrology, the study of how water moves through the environment and where it flows on the ground—crucial for detecting potential flood dangers. It also means evaluating the vulnerability of the population for factors like income, food security, and access to health care.

Ultimately, longer-range forecasts enable longer-range thinking, says Stephens, who also serves as the science lead at the Red Cross Red Crescent Climate Centre, an international group focused on how to address the impact of climate change and extreme weather on vulnerable people and communities. Instead of limiting action to a last-minute evacuation of people in a typhoon's path, forecast-based programs can encourage early harvesting of crops that would otherwise be wiped out and fixing roofs to minimize home damage.

Climate change only intensifies these needs, Stephens tells me, as we experience more unprecedented weather events, such as rapidly intensifying hurricanes and extreme floods. "We know that there are going to be events in the future that have never been seen before," she says. "But that doesn't mean that we can't prepare for them."

Conclusion

DOING EVEN MORE ABOUT THE WEATHER

As I near the end of my travels through the interconnected organizations and businesses of the weather-forecasting world, I find myself in Albany, New York, standing in a windowless room with Nick Bassill, a forty-something Ph.D. in meteorology. The room gives off serious start-up energy, immediately familiar to me from my days covering small companies at the beginning of the dot-com era. The furniture is simple and utilitarian: Bench-style desks line three walls, supporting a few computer workstations with room for several more. A wall-size computer display, showing a log-in screen and still partly shrouded in plastic wrapping, sits propped against a desk, waiting to be hung. The air gives off that new-office smell.

What I'm seeing—even though it's still ramping up—is a new and focused effort to do better at doing something about the weather. This is an operations space for New York's fledgling State Weather Risk Communication Center (SWRCC) and Bassill is its newly appointed director. Just a few weeks earlier, New York governor Kathy Hochul announced the launch of the SWRCC, describing it as a tool to protect the state "as we face the rising risk of extreme weather events."

Based at the University at Albany, the center operates as an information hub connecting weather experts, emergency managers, and officials throughout the state, from highway agencies to school districts. A key part of the center's mission can be found right there in the name:

communications. Even though Bassill is a meteorologist, he isn't here to forecast the weather so much as to explain those forecasts—to tell the story of the weather to come.

"Communication is one of the things that meteorologists are not necessarily great at," Bassill says. "We're really good at making a map, and we can understand the data and make a forecast. But there's also the presentation of that. How do you design something that an emergency manager could just look at and immediately get everything they need from it without being overwhelmed? It's really important when people are making decisions and they only have a minute to look at it and synthesize."

I wanted to see the SWRCC in person because it speaks to so many of the themes that have run through my reporting about weather forecasting and how to protect people. Of course, we must continue to improve the accuracy of predictions. But progress also depends on explaining those forecasts and making sure that anyone who needs to factor weather into a decision understands what's at stake. In that sense, Bassill and his new team aren't just communicators; they're translators. They can take their understanding of meteorology, use it to analyze forecasts, then pass on the important information to decision-makers in whatever form is most useful to them, improving the weather literacy of those decision-makers along the way.

Two members of the SWRCC team, Allison Finch and Trey Ryan, show me some of the tools they're developing. One is a report for the state Division of Homeland Security and Emergency Services. They open up a PDF file headed "Daily Weather Brief" that lists some bullet points atop a spreadsheet-like grid that has rows labeled with specific weather threats: rain, snow, wind, flooding, and a few others. The columns on this grid are days of the week, looking out seven days in advance. The boxes in the grid are left blank or filled in with colors according to a simple "impact scale classification" system: green for "minor," yellow for "moderate," orange for "major," and red for "extreme." An additional row specifies whether the forecast confidence level is low, medium, or high. Even without reading the bullet points, anyone could scan today's report and ascertain that there's a moderate concern about rain and flooding for the next day.

"These agencies are dealing with so many things," says Finch, a young University at Albany meteorology graduate who did a few stints in broadcasting before returning here to join Bassill's group. "We don't want them bogged down in their email. We want something they can look at and say, 'I don't see any yellow, I don't see any orange, I don't see any red.'" Recipients who want to dive deeper can click buttons on the front page of the PDF to access details. Finch and Ryan are quick to note that the format is a work in progress; they expect to adjust it as they get feedback from their end users.

The University at Albany, part of the State University of New York system, offers a logical home for this new effort. We're in ETEC, a gleaming building that opened in 2021. Much like the National Weather Center facility at the University of Oklahoma, ETEC brings together an assortment of meteorology-related organizations, both academic and governmental, under one roof. There's the university's own Department of Atmospheric and Environmental Sciences and its associated research center. The National Weather Service has based its Albany forecast office, with responsibility for eastern upstate New York along with parts of Connecticut, Massachusetts, and Vermont, in ETEC as well. Yet another space in the building houses the headquarters of the New York State Mesonet, which operates 126 weather stations to monitor local conditions. And, significantly, the university's program in emergency preparedness, homeland security, and cybersecurity also operates out of the building, putting the people training to respond to crises together with the weather experts. Albany's status as New York's capital situates ETEC near the offices of numerous state agencies with an interest in weather, such as the Department of Transportation.

Put it all together and you get a picture of how weather forecasting can integrate with the spectrum of official agencies that play a role in keeping the public safe. It's especially notable because, in the past few years, residents of New York City have suffered when leaders didn't pay enough attention to the weather or didn't understand the significance of the forecasts.

On September 1, 2021, the remnants of Hurricane Ida swept toward the Northeast and the Mid-Atlantic states, bringing record-setting rainfall and flash floods. The weather claimed forty-nine lives across the region.[191] Thirteen of those deaths were in New York City, most of them drownings that occurred when people weren't able to escape unregulated basement apartments.[192] The tragedy raised a host of questions about New York City's preparedness for extreme weather. But at a press conference the next day, Mayor Bill de Blasio suggested that lack of notice was a factor. "We're getting from the very best experts, projections that then are made a mockery of in a matter of minutes," de Blasio told the assembled reporters.[193] He added that the deluge "turned into the biggest single hour of rainfall in New York City history with almost no warning."

Watching the press conference on television, I was stunned. De Blasio seemed to be blaming the weather forecasts. But anyone who had been paying attention could have identified the risks building at least a day in advance. I had been watching those forecasts myself. Two days before the rain, the National Weather Service predicted a 20 percent to 50 percent chance of flash-flood rainfall in the New York City area for September 1. One day before the storm, forecasters upped their predictions and the urgency: "Considerable to significant and potentially life-threatening flash flooding is expected." Ultimately the rainfall amounts and rates were even worse than those forecasts—but the suggestion of "almost no warning" from the mayor sounded like a dodge.

In the aftermath, the city launched inquiries and plans for more effective responses in the future. That work was put to the test only two years later. On September 29, 2023, fragments of Tropical Storm Ophelia delivered a round of hammering rain. The storm effectively shut down the city, producing scenes of kids trapped in schools that had been flooded and city buses stuck with water rising into the passenger spaces. Fortunately, there were no deaths. But New York City once again appeared to be caught flat-footed, prompting comptroller Brad Lander to conduct an inquiry. Though he found some improvements over the Ida response, his report also noted some head-scratching lapses. Mayor Eric Adams, who had succeeded de Blasio, "had not appointed an Extreme Weather Coordinator at the time of the storm"—more than a year and a half into his term and even though the job had been established by de Blasio because of the Ida fiasco.[194]

In the context of these extreme events, efforts like the new State Weather Risk Communication Center in Albany offer a model that can hopefully improve responses by driving home the relevant information. (It should be noted that, in contrast with New York City's

Conclusion

Ophelia response, the state government acted with alacrity; Governor Kathy Hochul spoke publicly about the potential for "havoc" the day before the rain.) More broadly, people in communities everywhere should insist that their leaders find innovative ways to weave weather intelligence more seamlessly into the agencies and organizations that keep people safe and help society run smoothly. With each new extreme event, we see the value of advance warning. Investing in forecast improvements will pay off. But that value can only be realized when people understand and act on the forecasts. We should do better—and we can.

From everything I have seen and learned, the future promises more meaningful progress in forecasting the weather. Superior computer models, new sources of observations, deeper scientific understanding of atmospheric physics—all will contribute to more accurate predictions on timescales from minutes and days to weeks and months. In Norman, Oklahoma, the storm-chasers are gathering data for insights into how tornadoes form, the better to improve warning times. In San Diego, a utility company's dedicated sensor network monitors the hillsides and valleys in an effort to prevent wildfires. In Reading, near London, scientists steadily improve their computer modeling to predict the weather accurately days in advance. And in Albany, New York, Bassill and his colleagues work to help decision-makers interpret what the forecasts are telling them.

All these efforts and more build on steady improvements over the past few decades that have made forecasts more accurate and more trustworthy, even if that headway isn't always recognized by the public. In years to come, meteorologists will tap sources of data from miniature satellites to everyday cell phones. Forecasts will get more prescient at hyperlocal scales and when looking months ahead. Artificial intelligence shows game-changing potential—but, as with every realm confronting the power of AI, this new technology must be introduced carefully and with appreciation for its drawbacks.

Perhaps even more encouraging, leaders and organizations throughout the weather enterprise are putting more focus on the point of all this work: protecting the people whose homes and livelihoods—and, sometimes, their very lives—depend on making the best decisions in the face of weather threats. The growing role of social scientists offers new opportunities to turn good forecasts into better outcomes. We need to keep up the pace and find the most effective ways to communicate so everyone can understand what to do in times of peril, regardless of where they live, how much money they have, or what language they speak.

For all this, we owe a debt to the cloud warriors. From my earliest conversations with meteorologists and researchers, I have been impressed over and over again with the ideals of duty and service that permeate their work. So many of the people I've introduced throughout this book are driven not only to expand knowledge out of scientific curiosity but also to use what they learn to defend against an often hostile environment. Their work may not always be perfect but they are highly aware of the stakes. They're out there pushing the frontiers every day without much recognition outside their professional and academic circles. Yet their contributions help protect property, expand economic opportunities, and, most of all, safeguard human lives.

But we need to expect—even demand—more progress. Why? For one thing, climate change keeps raising the stakes. The warming of the global climate brings sobering consequences, from hurricanes that dump more flood-producing rains to higher risks from wildfires. The news just keeps coming. One recent study from climate scientists concluded that for the twelve-month period ended May 15, 2024, "human-caused climate change added an average of 26 days of extreme heat" on average across the world.[195] We also continue to learn more about the downstream economic repercussions. For instance, among other findings, the US government's Fifth National Climate Assessment in late 2023 reported that heavy rains and flooding have hurt corn-crop yields to an extent comparable to effects from droughts.[196]

To keep moving forward, several areas bear watching. One is support for NOAA and the National Weather Service in the second Trump administration. When Donald Trump first became president in 2017, his administration sought to slash NOAA's budget by 16 percent, though the spending ultimately approved by Congress avoided cuts and gave a slight increase.[197] When President Joe Biden took office, his administration requested a record increase in spending, citing climate and weather needs. Though the proposed jump of more than 28 percent got scaled back, NOAA still received an 8 percent increase. By fiscal 2024, the total approved NOAA budget was $6.3 billion, a sharp contrast with the $4.8 billion level that Trump wanted but didn't get for fiscal 2018. Of the 2024 NOAA budget, the National Weather Service got $1.35 billion. Of course, every government agency should be scrutinized for waste and inefficiency, but considering the effects of extreme weather, NOAA's funding should reflect the growing economic costs. For 2023, the United States suffered twenty-eight weather and climate disasters, according to NOAA, with damages of at least $94 billion.[198]

Over a longer term, there are bigger investments to consider. For example, the nation's current network of NEXRAD weather radars dates back to 1988. While those radars have received upgrades over the years, they don't thoroughly cover the entire United States, leaving gaps that make it harder for meteorologists to detect severe weather. A newer technology called phased-array radar could substantially enhance tornado and thunderstorm warnings by scanning more quickly and allowing meteorologists to zoom in on specific danger areas. NOAA says NEXRAD will operate at least through 2035, and a phased-array system is among the options that would be considered after that. But it's far from clear that legislators will fund such improvements.

Future forecasting budget battles threaten to play out in a dangerously politicized environment as the United States struggles with polarization and extremist views. Based on recent history, some right-wing voices can be expected to push for more privatization and a narrowing of the mission of the National Weather Service. The Project 2025 group called for NOAA to be broken up, complaining of "climate alarmism." It's a shocking dismissal of the agency's scientific, national-security, and public-safety value. Time will tell how aggressive Trump will be in his second term when it comes to scaling back NOAA or NWS—and whether legislators, and their constituents who depend on forecasting, will push back. (Even AccuWeather, which has advocated in the past for the Weather Service to cede more duties to the private sector, said it didn't support the Project 2025 recommendation.) Blowing up the weather-enterprise balance seems like folly. There's unquestionably a role for private innovation. But when it comes to forecasts with lives on the line, the public deserves the accountability of elected officials that comes from government services.

Just as politicization in weather mirrors broader divisions in the United States, some other important issues track with contentious societal debates. Good communication is vital to weather safety, and social science researchers continue to learn about more effective ways to convey warnings. Yet thanks largely to the rise of social media, the channels through which many people consume information keep shifting and fragmenting, at times in unsettling ways. Those who watch traditional television news can still get their updates from trusted meteorologists. But the world of Facebook, Twitter (now X), and TikTok presents a jumble of reliable alerts, uninformed speculation, and pure misinformation. One indicator of this shift: A Pew Research survey found that 43 percent of TikTok users say they regularly get news from the social media network.

When used thoughtfully, social media can provide reliable and timely updates and direct links to official information. I often turned to the excellent Twitter feeds from individual National Weather Service local forecast offices. But after Elon Musk purchased Twitter, rebranded it as X, and made a slew of controversial changes, the network saw users drift away, with some estimates putting the decline in daily users of the app at 23 percent over nearly a year and a half.[199] These kinds of audience shifts make it tricky for officials and researchers to commit to a platform. How long can the Weather Service justify spending resources on X if its audience isn't stable? As a user who benefited from following numerous dependable weather experts on Twitter for years, I've been distressed to watch the community decline. And casual users who don't go directly to quality sources but drift at the mercy of the algorithms can easily get misled. In one example reported by *The New York Times*, in early 2024 meteorologists and California officials had to reassure the public after a viral social media post from an actress suggested the state was at risk for a mega-storm that would bring disastrous flooding.[200]

Progress is also needed in another divisive area: addressing inequities. From historically redlined Black neighborhoods that experience disproportionate effects of extreme heat to underserved communities such as the Amish, who need ways to access information that are compatible with their beliefs, to low-income individuals forced to rent illegal basement apartments like the ones that flooded in New York City, I have seen how race, class, and other factors put minority groups at particular risk. Another notable example are tribal communities, some of which are unusually vulnerable to heat, wildfires, and floods due to their rural locations and lack of resources. These needs underscore the role for local Weather Service offices to strengthen relationships with community leaders and devise specific plans to enhance preparation.

Inequities play out on a global scale as well. Developing nations typically don't have the robust infrastructure of weather sensors and radars needed for accurate forecasting. They may also lack the resources to buy and operate supercomputers necessary for numerical weather models. Yet, just as with the agricultural communities in Ghana and Zimbabwe that I examined, capable forecasting can make a tremendous difference in people's lives. The good news: Artificial intelligence shows the potential to offer potent forecasting in these places at lower costs, since the AI systems typically don't require the computing power necessary for traditional models. Particularly as climate change shapes the weather, we must ensure that, when it comes to critical forecasting, no place gets left behind.

The way forward requires putting everything together: scientific research, cutting-edge technology, communication optimized from behavioral insights, and solutions for obstacles within specific populations and communities. Progress demands long-term adaptation, such as ensuring that suitable tornado shelters are available to people living in mobile homes, beefing up infrastructure like drains for flood-prone areas, and implementing rules to protect outdoor workers from extreme temperatures. It also requires getting ready to react nimbly in the short term to forecasts of specific events—preparing to evacuate from a hurricane's path when recommended or laying in supplies ahead of a winter storm that could knock out power.

Everyone should want to become weather-literate and look for ways to counteract the threats. Government officials at the local, state, and national levels need to ask whether they have prepared their organizations. Executives and managers in the corporate world can look for ways to integrate weather intelligence into their operations, keeping workers safe and identifying business opportunities. No matter your role, you can think about how weather will affect what you do and the people you interact with.

We can also act smarter as individuals. It's as easy as paying closer attention to the forecast and thinking through plans for bad weather in all circumstances. When you're more weather-literate, you'll grasp just how dangerous flash floods can be and avoid driving into water. You'll take note of a tornado watch and figure out where you will shelter if the warning of an actual twister arrives. You'll also better appreciate that in a changing climate, past experience isn't a sufficient guide to the future, as residents in the Pacific Northwest found out during the 2021 heat wave. We need to get ready for all kinds of weather.

As I write these last words, I marvel at how many ways I have to tap into information about the weather. My smartphone has multiple apps that summarize forecasts generated from mountains of data on massive computers. It monitors my own personal weather station to give me a heads-up on nearby lightning. It receives automated text alerts from my city to warn me of weather emergencies. I can call up the latest radar images, read what National Weather Service forecasters have to say, and learn from meteorology experts on social media—all from a device in my pocket. These feeds are powered by predictions that are more accurate than ever, thanks to the cloud warriors pushing the frontiers of forecasting. In short, there has never been a time when it was easier to go beyond just talking about the weather to doing something about it.

Acknowledgments

All journalists know—even as we put our names on the covers of books or the bylines of articles—how much our work depends on the willingness of others to share their knowledge and experience. In the case of *Cloud Warriors*, those contributors number in the hundreds: people throughout the weather enterprise who took time to discuss their work, reveal their hopes, express their concerns, and pull back the curtain on this vital undertaking. One of the great rewards of being a journalist is getting what amounts to a backstage pass to the events that shape our world. The opportunities I had to go behind the scenes in forecasting have been a privilege.

To everyone who spoke with me for this project, you have my deep thanks. Many of their names are found in the pages of this book, but many others are not—as I was reminded when I sifted through my notebooks and interviews, and realized the abundance of people and insights I couldn't fit into my manuscript. Regardless, I hope this book helps readers appreciate just how much brainpower and hard work go into the forecasts on which we depend.

This book would not have been possible without the guidance and support of my remarkable literary agent, David Halpern. From the start, he shared my conviction that, even amid so much attention on climate change, the need for progress in weather forecasting was a largely untold story. David wore many hats: a tutor for this first-time author on the publishing landscape; an editor whose sharp pencil greatly improved my proposal; and a therapist/counselor willing to listen with gentle encouragement and the occasional firm nudge. I'm also grateful to my onetime *Newsweek* colleague Mike Giglio for recommending David. As I recall Mike's conversation over hot dogs at a Mets game, he told me: "David gets journalists." Lucky for me, he really does.

At St. Martin's Press, I'm indebted to my talented and considerate editor, Anna deVries. Her insightful questions and suggestions made her an invaluable collaborator. My thanks also to the entire St. Martin's team, including managing editor Lizz Blaise and associate publisher Laura Clark. Soleil Paz designed a cover that elegantly channeled the book's focus. To Gabrielle Gantz and Michelle Cashman, who led publicity and marketing, respectively, my gratitude for all your efforts to get out the word. Special thanks to Daniela Rapp for her pivotal early interest. Tracy Roe, who copyedited the manuscript, brought an eagle eye for needed fixes and a deft hand in massaging out some of the more distracting tics of my writing. My astute researcher, Sara Krolewski, was relentless in fact-checking my manuscript.

At many points while reporting and writing this book, I found myself reflecting on how fortunate I've been to learn from so many colleagues and friends over the years. From my time at *The Wall Street Journal*, that list includes Larry Rout, Melinda Beck, Dennis Kneale, Jonathan Dahl, and Amy Stevens. It also includes the leadership from that era at the *Journal*—namely, Paul Steiger, Dan Hertzberg, and Barney Calame—giants who set standards and supported their journalists with a dedication that I have always sought to live up to. Jane Berentson showed me how to approach serious work without taking yourself too seriously. Thank you to Paula Szuchman for some helpful early conversations and for prodding me to follow my interest in forecasting. My friend Becky Quick has taught me the art of asking the right questions (which, often, means asking all of them). Joanne Lipman always has the best advice just when it's needed.

At *Time*, I was surrounded by a team that offered a daily master class in storytelling. Rick Stengel, Nancy Gibbs, Edward Felsenthal, and Sam Jacobs have each led *Time* with a vision that honors the magazine's legacy and its commitment to readers. Thank you also to Michael Duffy, Chrissy Dunleavy, Radhika Jones, Ratu Kamlani, Paul Moakley, D. W. Pine, and

ACKNOWLEDGMENTS

Kira Pollack for good advice and treasured camaraderie over the years. Ben Goldberger and Bryan Walsh were essential sounding boards for me getting this project off the ground.

I also want to highlight a debt to the many journalists I've never met but whose work informed my research—particularly the irreplaceable service performed by those who cover local news. From tornadoes to heat waves to wildfires, their immediate on-the-ground reporting captures an essential record of the impact from severe weather. Even though local news has been buffeted by financial tides and digital upheavals, we need this journalism more than ever.

On days when that blank computer screen consumed my field of vision, I often found inspiration in talking with young journalists. It has been a pleasure to work as an adjunct with talented students at Columbia University's Graduate School of Journalism. Seeing the gusto with which they approached ambitious projects helped put my own work in perspective. Talking with student journalists at Princeton University, where I serve as a trustee of *The Daily Princetonian*, has been similarly enriching. Many years later, I still turn to my own friends from college days at the *Princetonian* for counsel, especially Laurence Hooper and Doug Widmann.

Most of all, my love and gratitude to my family for their unwavering support. My wife, Tracey Weber, has been there with me through writer's block and deadline stress going all the way back to the newspaper at our high school. When I realized that I wanted to leave newsroom management behind and return to writing and tackle this book, she encouraged me without hesitation. She is my partner and my light. Much love and thanks also to our daughters, Abby and Ellie, for cheerfully tolerating those times when work collided with family plans and for helping me see the world through younger eyes—even now that they are adults themselves. I am proud of you every day. Thanks also to my parents, Jim and Sally Weber, who encouraged my early fascination with science and technology—a gift that has nourished my work ever since.

Finally, gratitude to you, the reader. Whether it's this book, or another book, or a news site, radio broadcast, television show, newsletter, or podcast, journalists are more appreciative of your interest and support than you can know. Thank you for reading.

Glossary

Advanced Weather Interactive Processing System (AWIPS)

Automated Surface Observing System (ASOS)

CAMALIOT (appli*ca*tion of *ma*chine *l*earning technology for GNSS *IoT* data fusion)

Climate Prediction Center (CPC)

Comprehensive Bespoke Atmospheric Model (CBAM)

Cooperative Institute for Severe and High-Impact Weather Research and Operations (CIWRO)

European Centre for Medium-Range Weather Forecasts (ECMWF)

Geophysical Fluid Dynamics Laboratory (GFDL)

Geostationary Lightning Mapper (GLM)

Geostationary Operational Environmental Satellites (GOES) program

Global Forecast System (GFS)

High-Resolution Rapid Refresh model (HRRR)

Integrated Forecasting System (IFS)

National Blend of Models (NBM)

National Center for Atmospheric Research (NCAR)

National Hurricane Center (NHC)

National Integrated Heat Health Information System (NIHHIS)

National Oceanic and Atmospheric Administration (NOAA)

National Severe Storms Laboratory (NSSL)

National Weather Center (NWC)

National Weather Service (NWS)

numerical weather prediction (NWP)

public-safety power shutoff (PSPS)

quasi-linear convective system (QLCS)

Santa Ana Wildfire Threat Index (SAWTI)

State Weather Risk Communication Center (SWRCC)

Storm Prediction Center (SPC)

Structural Extreme Events Reconnaissance Network (StEER)

Unified Forecast System (UFS)

Verification of the Origins of Rotation in Tornadoes Experiment (VORTEX)

Warn-on-Forecast System (WoFS)

weather forecast office (WFO)

wildland-urban interface (WUI)

Notes

INTRODUCTION

1. I'm indebted here to the diligent efforts of the researcher behind the Quote Investigator website, Garson O'Toole (a pseudonym). Some of the confusion stems from a version of the quote that appeared in 1897 in the *Hartford Courant*, of which Warner was an editor. The paper described its version as being something "a well known American writer said once," suggesting to some that Warner was referring to Twain. Quote Investigator reports turning up earlier instances definitely attributable to Warner, such as an item in 1884. For more details, see https://quoteinvestigator.com/2010/04/23/everybody-talks-about-the-weather.

2. *California's Fourth Climate Change Assessment: Statewide Summary Report*, State of California, 2018, https://www.energy.ca.gov/sites/default/files/2019-11/Statewide_Reports-SUM-CCCA4-2018-013_Statewide_Summary_Report_ADA.pdf.

3. For additional details, see the chapter on hurricanes, but a good summary of the research regarding hurricane intensity can be found in Thomas Knutson et al., "Tropical Cyclones and Climate Change Assessment: Part II: Projected Response to Anthropogenic Warming," *Bulletin of the American Meteorological Society* 101, no. 3 (March 2020): E303-22, https://doi.org/10.1175/BAMS-D-18-0194.1.

4. I won't attempt to recommend an exhaustive list of climate-focused books, but there are many excellent treatments out there, among them Elizabeth Kolbert's *Field Notes from a Catastrophe: Man, Nature, and Climate Change* and Jeff Goodell's *The Heat Will Kill You First: Life and Death on a Scorched Planet*.

5. It's a true triumph of science and one that deserves more recognition. See Peter Bauer, Alan Thorpe, and Gilbert Brunet, "The Quiet Revolution of Numerical Weather Prediction," *Nature* 525, no. 7567 (September 2015): 47-55, https://doi.org/10.1038/nature14956, and Richard B. Alley, Kerry A. Emanuel, and Fuqing Zhang, "Advances in Weather Prediction," *Science* 363, no. 6425 (January 2019): 342-44, https://doi.org/10.1126/science.aav7274.

6. "National Hurricane Center Forecast Verification," National Hurricane Center, last updated June 7, 2024, https://www.nhc.noaa.gov/verification/verify5.shtml.

7. We're familiar these days with different varieties of literacy, such as financial literacy, digital literacy, and media literacy. So far, the phrase *weather literacy* hasn't attained the same standing, but it's a helpful framework. A trio of researchers in Germany conducted a survey to gauge weather literacy among the public in the context of climate change and extreme weather. In their 2019 paper, they wrote that the "deficits in weather literacy" they found indicated the need to better explain the impact of dangerous weather and communicate forecast uncertainty—key concepts for many of the experts I've talked with. See Nadine Fleischhut, Stefan M. Herzog, and Ralph Hertwig, "Weather Literacy in Times of Climate Change," *Weather, Climate, and Society* 12, no. 3 (July 1, 2020): 435–52, https://doi.org/10.1175/WCAS-D-19-0043.1.

8. The conservative group Project 2025 detailed these plans in "Mandate for Leadership: The Conservative Promise," a document described as a "comprehensive, concrete transition plan for each federal agency." See https://www.project2025.org/playbook/.

9. The title of the exhibit also demonstrates the winding path through culture that Charles Dudley Warner's aphorism has taken over the decades. Curators indicated that the exhibit title referenced a landmark 1968 poster from the Socialist German Student Union that proclaimed "Alle reden vom Wetter. Wir nicht" ("Everybody talks about the weather. We don't") alongside images of Marx, Engels, and Lenin—a suggestion that socialists had more important things to do than discuss the rain. (The socialist group's poster is itself considered to have referenced an earlier West German rail poster that used the same slogan to indicate that the railway's trains ran on time regardless of the weather.) More recently, the exhibit curators noted, German artist Anne-Christine Klarmann took the concept but used images of Greta Thunberg and other environmental activists with the tweaked slogan "Everybody talks about the weather. So do we"—a formulation that emphasizes the need for action.

1. TORNADOES

10. The prefix *meso* comes from the Greek word *mesos*, "middle." Larger occurrences, such as hurricanes and weather fronts, are known as synoptic-scale phenomena, from the Greek *súnopsis* and Latin *synopsis*, referring to a whole or general view.

11. Joshua Wurman et al., "Low-Level Winds in Tornadoes and Potential Catastrophic Tornado Impacts in Urban Areas," *Bulletin of the American Meteorological Society* 88, no. 1 (January 2007): 31–46, https://doi.org/10.1175/BAMS-88-1-31. Note that the authors discuss wind speeds in meters per second rather than miles per hour.

12. The number 158 reflects direct fatalities. Including indirect fatalities, the Joplin death toll stands at 161. Damage estimates vary but in the years since, $2.8 billion has become the most frequently cited figure and has been used by the National Institute of Standards and Technology. See Jonathan Griffin, "The Joplin Tornado: A Calamity and a Boon to Resilience, 10 Years On," National Institute of Standards and Technology, May 21, 2021, https://www.nist.gov/feature-stories/joplin-tornado-calamity-and-boon-resilience-10-years.

13. The annual mean warning lead time for all tornadoes ranged from eight to ten minutes from 2014 through 2023, according to figures provided to me by the National Weather Service. That averages out to nine minutes. Average warning times in the preceding five years were a few minutes longer, but the ratio of false alarms to actual tornadoes was also greater, illustrating the tension between lowering the number of false alarms and extending warning lead times.

14. The average annual false-alarm ratio ranged from 66 percent to 72 percent from 2014 through 2023. In the five years prior, the annual average ranged from 72 percent to 77 percent.

15. From 2014 through 2023, if only EF2 to EF5 tornadoes are considered, annual average warning lead times ranged from twelve to seventeen minutes, according to National Weather Service figures.

16. Weather Research and Forecasting Innovation Act of 2017, Pub. L. No. 115–25, 131 Stat. (2017).

17. C. Donald Ahrens and Robert Henson, *Meteorology Today: An Introduction to Weather, Climate, and the Environment* (Boston: Cengage Learning, 2021).

18. If you have a newer oven, perhaps it has a convection mode—though that's something of a misnomer. Convection takes place anytime there's a difference in temperature between two air masses. A more accurate term for a convection oven would be *forced convection*. Normally, the warm air next to the food being cooked will cool as the colder food absorbs some heat. A convection oven uses a fan to force warmer air to move around and displace that cooled air, thereby heating the food more efficiently.

19. Tetsuya Theodore Fujita, *Memoirs of an Effort to Unlock the Mystery of Severe Storms During the 50 Years, 1942–1992* (Chicago: Wind Research Laboratory, 1994).

20. Roscoe R. Braham, "The Thunderstorm Project 18th Conference on Severe Local Storms Luncheon Speech," *Bulletin of the American Meteorological Society* 77, no. 8 (August 1, 1996): 1835–46, https://doi.org/10.1175/1520-0477-77.8.1835.

21. "Enhanced F Scale for Tornado Damage," Storm Prediction Center, National Weather Service, accessed April 14, 2024, https://www.spc.noaa.gov/efscale/ef-scale.html.

22. Kelsey J. Mulder and David M. Schultz, "Climatology, Storm Morphologies, and Environments of Tornadoes in the British Isles: 1980–2012," *Monthly Weather Review* 143, no. 6 (June 2015): 2224–40, https://doi.org/10.1175/MWR-D-14-00299.1.

23. Jonathan M. Davies and Anthony Fischer, "Environmental Characteristics Associated with Nighttime Tornadoes," *Electronic Journal of Operational Meteorology* (2009), http://nwafiles.nwas.org/ej/pdf/2009-EJ3.pdf.

24. Ernest Agee et al., "Spatial Redistribution of U.S. Tornado Activity Between 1954 and 2013," *Journal of Applied Meteorology and Climatology* 55, no. 8 (August 2016): 1681–97, https://doi.org/10.1175/JAMC-D-15-0342.1.

25. "About the National Weather Center," University of Oklahoma, accessed April 14, 2024, http://www.ou.edu/nwc/partners.html.

26. "TOTO Home Page (Online Tornado FAQ)," Storm Prediction Center, accessed April 14, 2024, https://www.spc.noaa.gov/faq/tornado/toto.htm.

27. It's usually illustrated with the example of the noise made by a train or a siren as it moves toward you, passes you, and moves away from you. The compressed audio wave sounds high-pitched as it moves toward you, and the pitch drops as it moves away.

28. Spotters are trained volunteers who alert the National Weather Service in real time when they observe severe weather. The NWS estimates there are somewhere between 350,000 and 400,000 of these volunteers across the United States submitting their observations to their local NWS offices. (If this sounds like fun, you can learn more about joining at weather.gov/skywarn.)

29. The data from these satellites is critical for the big global computer weather models. For more about them, see chapter 6.

30. Accounts from eyewitnesses were gathered by a team from the National Institute of Standards and Technology; the full report makes for a sober read. See *Final Report, National Institute of Standards and Technology Technical Investigation of the May 22, 2011, Tornado in Joplin, Missouri*, National Institute of Standards and Technology, March 26, 2014, https://doi.org/10.6028/NIST.NCSTAR.3.

31. "NWS Central Region Service Assessment: Joplin, Missouri, Tornado—May 22, 2011," National Oceanic and Atmospheric Administration, July 2011, https://www.weather.gov/media/publications/assessments/Joplin_tornado.pdf.

32. VORTEX-SE Scientific Steering Committee, *VORTEX-SE Science Assessment*, National Severe Storms Laboratory, September 2020, https://inside.nssl.noaa.gov/vsecommunity/wp-content/uploads/sites/34/2020/11/Science_Assessment_Sep2020_formatted_V2.pdf.

33. "Beauregard-Smiths Station Tornado—March 3, 2019," National Weather Service, accessed April 14, 2024, https://www.weather.gov/bmx/event_03032019beauregard.

34. The SPC outlooks aren't generally considered a tool for the public; they're aimed at giving local forecasters a heads-up. The work of translating the risk from an SPC outlook tends to fall on the local Weather Service office as well as television and radio meteorologists. It was only after my visit to NSSL that I began to follow the SPC forecasts directly.

35. StEER is a National Science Foundation–supported group dedicated to analyzing the effect of natural disasters on buildings. It deployed field assessment teams with researchers from Auburn University and the University of South Alabama to survey the damaged areas; see D. Roueche et al., *March 3, 2019 Tornadoes in Southeast United States: Early Access Reconnaissance Report*, Structural Extreme Events Reconnaissance Network, June 4, 2019, https://www.designsafe-ci.org/data/browser/public/designsafe.storage.published/PRJ-2265.

36. "Tornado Event at Elk City, OK, May 16, 2017," National Severe Storms Laboratory, May 16, 2017, https://www.nssl.noaa.gov/projects/wof/casestudies/elkcity-16may2017.

2. FIRE

37. The Calscape database from the California Native Plant Society offers a rich repository of information for gardeners or anyone looking to better understand the state's flora; you can find it at https://calscape.org.

38. Another deep repository of plant information can be found at the Fire Effects Information System database from the US Forest Service. It details how specific species interact with and are affected by fire. Find it online at https://www.feis-crs.org/feis/.

39. In California, wildfires are typically given names by fire dispatchers or officials at the outset, usually based on their location. The practice helps simplify communications, especially when multiple fires are underway. The Cedar Fire took its name from the Cedar Creek area of the Cleveland National Forest.

40. "2003—Cedar Fire," City of San Diego, accessed May 2, 2024, https://www.sandiego.gov/fire/about/majorfires/2003cedar.

41. *Valley Fire After Action Report,* County of San Diego, September 5, 2020, https://www.alertsandiego.org/content/dam/alertsandiego/preparedness/en/aar/Valley%20Fire%20AAR%20Final%20120120.pdf.

42. Volker C. Radeloff et al., "Rapid Growth of the US Wildland-Urban Interface Raises Wildfire Risk," *Proceedings of the National Academy of Sciences* 115, no. 13 (March 2018): 3314–19, https://doi.org/10.1073/pnas.1718850115.

43. See the comments of Natasha Stavros, at the time a science systems engineer at NASA's Jet Propulsion Laboratory. (Given its role in Earth observations, NASA conducts extensive climate research.) Alan Buis, "The Climate Connections of a Record Fire Year in the U.S. West," NASA Science, last updated March 18, 2024, https://science.nasa.gov/earth/climate-change/the-climate-connections-of-a-record-fire-year-in-the-us-west/.

44. Cal Fire keeps updated lists of the most destructive and most deadly fires; see https://www.fire.ca.gov/our-impact/statistics.

45. See "NOAA National Centers for Environmental Information (NCEI) U.S. Billion-Dollar Weather and Climate Disasters," National Centers for Environmental Information, https://www.ncei.noaa.gov/access/billions/. NCEI updates these damage figures to current dollars with an adjustment based on the consumer price index. All NCEI damage-cost figures used in this book reflect the CPI-adjusted numbers as of June 2024.

46. In recounting events of the Camp Fire, I have relied heavily on the public report of the Butte County district attorney's office, which undertook a sizable investigation that led to the criminal indictment to which PG&E later pleaded guilty; see *The Camp Fire Public Report: A Summary of the Camp Fire Investigation,* Butte County District Attorney's Office, June 16, 2020, https://www.buttecounty.net/DocumentCenter/View/1881/Camp-Fire-Public-Report—Summary-of-the-Camp-Fire-Investigation-PDF.

47. *SED Camp Fire Investigation Report*, Safety and Enforcement Division, California Public Utilities Commission, November 8, 2019, https://www.cpuc.ca.gov/industries-and-topics/wildfires/wildfires-staff-investigations.

48. "PG&E Statement on Company's Guilty Plea Related to 2018 Camp Fire," PG&E Corporation, June 16, 2020, https://investor.pgecorp.com/news-events/press-releases/press-release-details/2020/PGE-Statement-on-Companys-Guilty-Plea-Related-to-2018-Camp-Fire/default.aspx. The number of involuntary manslaughter counts is one less than the authorities' death toll of eighty-five; it excludes a casualty who was deemed not to have died as a direct result of the fire. See J. D. Morris and Lizzie Johnson, "Drama Marks PG&E Pleas," *San Francisco Chronicle,* June 17, 2020.

49. PG&E had already been under scrutiny for previous lapses, such as the 2010 San Bruno explosion. In that tragedy, a PG&E natural gas pipeline running through a residential area south of downtown San Francisco and near the city's airport blew up, razing nearby homes and killing eight people.

50. *Incident Investigation Report,* Safety and Enforcement Division, California Public Utilities Commission, October 9, 2023, https://www.cpuc.ca.gov/-/media/cpuc-website/divisions/safety-and-enforcement-division/investigations-wildfires/dixie-fire-investigation-report.pdf.

51. "2007—Witch Creek and Guejito Fires," City of San Diego, accessed May 2, 2024, https://www.sandiego.gov/fire/about/majorfires/2007witchcreek.

52. *Report of the Consumer Protection and Safety Division Regarding the Guejito, Witch and Rice Fires,* Consumer Protection and Safety Division, California Public Utilities Commission, September 2, 2008, https://docs.cpuc.ca.gov/PUBLISHED/FINAL_DECISION/93739-08.htm.

53. "A History of Significant Weather Events in Southern California: Organized by Weather Type," National Weather Service, March 2024, https://www.weather.gov/media/sgx/documents/weatherhistory.pdf.

54. Summer Lin, "Cigarette Butt Sparked Fire That Burned More Than 300 Acres, California Officials Say," *Sacramento Bee,* October 27, 2020.

55. "NWS San Diego All-Hazard Reference Guide," San Diego Weather Forecast Office, National Weather Service, August 2015, https://www.weather.gov/media/sgx/documents/WWA_Criteria.pdf.

56. Mark V. Thornton, "The History of Cal Fire," California Department of Forestry and Fire Protection, 1995, https://www.fire.ca.gov/about/our-organization.

57. The inmate program inspired a 2022 television series, *Fire Country,* on CBS. In real life, the program has been controversial at times. Some former inmates have described positive experiences as firefighters, but given the instances where participants have suffered injury and even death, critics have faulted the program as exploitative.

58. Leininger is also trained in geographic information systems, or GIS, an acronym that pops up constantly these days, not only in weather applications but in everything from pandemic monitoring to agriculture to retail—really, any endeavor that can benefit from seeing data mapped out to spaces in the physical world.

59. *Spreading Like Wildfire: The Rising Threat of Extraordinary Landscape Fires,* United

Nations Environment Programme, February 23, 2022, https://www.unep.org/resources/report/spreading-wildfire-rising-threat-extraordinary-landscape-fires.

60. The prediction was included in *California's Fourth Climate Change Assessment*.

61. "RAL's Super-Efficient, GPU-Enabled Microscale Model FastEddy Is Now Open Source!," Research Applications Laboratory, National Center for Atmospheric Research, accessed May 3, 2024, https://ral.ucar.edu/news/rals-super-efficient-gpu-enabled-microscale-model-fasteddyr-now-open-source. The FastEddy model runs on an NCAR computer cluster that uses GPU chips from NVIDIA, the company that makes graphics cards for a variety of brand-name personal computers. NVIDIA has evolved into an influential company in supercomputing and artificial intelligence.

62. Not incidentally, it also promises to model the complex movement of air around buildings in urban centers, a boon to drone companies planning to operate in cities.

3. THE LOCAL FORECAST

63. If you're curious about which office handles your local weather, see this online map: https://www.weather.gov/srh/nwsoffices.

64. Lans P. Rothfusz, "StormReady: From Idea to National Program," accessed April 25, 2024, https://www.weather.gov/stormready/history.

65. "National Weather Service Instruction 10–1802: The StormReady Recognition Program," National Weather Service, October 25, 2023, https://www.nws.noaa.gov/directives/sym/pd01005003curr.pdf.

66. "NOAA's National Weather Service Strategic Plan: Building a Weather-Ready Nation," National Weather Service, June 2011, https://www.weather.gov/media/wrn/strategic_plan.pdf.

67. The State College office is also responsible for issuing forecasts for seven airports in the area, a duty that typically gets handled by either the short-term or long-range forecaster.

68. Since my visit, Banghoff has been promoted to lead meteorologist, which has altered his rotation schedule a bit.

69. You can see a full list of NWS products, ranging from the Weather Roundup to the Monthly Hydrometeorological Plain Language Product, here: https://forecast.weather.gov/product_types.php?site=NWS.

70. To prepare the predictions they share with the public, broadcast meteorologists may draw on some of the raw data that the WFO forecasters used, take the WFO forecast as a starting point and apply their own knowledge of the area's quirks, or simply relay the WFO predictions in layperson's terms.

71. "Amish Population, 2023," Young Center for Anabaptist and Pietist Studies, Elizabethtown College, 2023, https://groups.etown.edu/amishstudies/statistics/population-2023/.

72. For more information on the WARN initiative, see "Weather Awareness for a Rural Nation (WARN): Developing Weather Safety Tools for Amish Communities," National Oceanic and Atmospheric Administration, January 31, 2023, https://www.noaa

.gov/education/stories/weather-awareness-for-rural-nation-warn-developing-weather-safety-tools-for-amish-communities. Also: "Initiative Helps Rural Amish Communities Become Weather Ready in Jackson, Kentucky," Federal Emergency Management Agency, March 29, 2024, https://www.fema.gov/case-study/initiative-helps-rural-amish-communities-become-weather-ready-jackson-kentucky.

73. Among the models included in the NBM are the GFS, the HRRR, non-US models from the European Centre for Medium-Range Weather Forecasts, Canada's forecasting center, and more. For full details see "Greater than the Sum of Its Parts . . . the NWS National Blend of Models," National Weather Service, accessed June 20, 2024, https://www.weather.gov/news/200318-nbm32.

74. For more on that incident, see Hannah Fingerhut, Heather Hollingsworth, and Summer Ballentine, "An Iowa Meteorologist Started Talking About Climate Change on Newscasts. Then Came the Harassment," Associated Press, July 8, 2023, https://apnews.com/article/meteorologist-harassment-threats-climate-change-iowa-bf91adbd26ca5e97507406947b47d684. The incident also prompted the American Meteorological Society to issue a statement decrying abuse of forecasters; see "Special Statement on Harassment and Intimidation of Broadcast Meteorologists," American Meteorological Society, June 30, 2023, https://www.ametsoc.org/index.cfm/ams/about-ams/ams-statements/statements-of-the-ams-in-force/special-statement-on-harassment-and-intimidation-of-broadcast-meteorologists/.

75. *Reauthorizing the Weather Act: Data and Innovation for Predictions, Before the House Committee on Science, Space and Technology, Subcommittee on Environment,* March 28, 2023 (statement of Antonio J. Busalacchi Jr., president, University Corporation for Atmospheric Research).

76. National Weather Service Duties Act of 2005, S. 786, 109th Cong. (2005).

77. Kimberly Hefling, "Santorum's Weather-Related Bill Criticized," Associated Press, May 27, 2005.

78. Michael Brice-Saddler, "A Trump Nominee's Family Company Paid $290,000 Fine for Sexual Harassment and Discrimination," *Washington Post,* February 12, 2019, https://www.washingtonpost.com/politics/2019/02/12/trump-nominees-family-company-paid-fine-sexual-harassment-discrimination/.

79. For a compelling exploration of the pressures to privatize more of the government's weather work, see *The Fifth Risk,* a 2018 book by Michael Lewis, the author of *Moneyball* and *Liar's Poker.* Its overall theme is the Trump administration's disdain for government expertise and institutional knowledge, and it includes a chapter on weather and climate.

80. You can read about Jefferson's weather-record habits and examine his observations at this digital repository: https://jefferson-weather-records.org/node/40573.

81. "Warmest and Coldest Days at Central Park (1869 to Present)," National Weather Service, accessed April 26, 2024, https://www.weather.gov/media/okx/Climate/CentralPark/warmcolddays.pdf.

4. HYPERLOCAL WEATHER

82. "GFS," National Centers for Environmental Prediction, National Oceanic and Atmospheric Administration, accessed May 5, 2024, https://www.emc.ncep.noaa.gov/emc/pages/numerical_forecast_systems/gfs.php.

83. "HRRR," National Centers for Environmental Prediction, National Oceanic and Atmospheric Administration, accessed May 5, 2024, https://emc.ncep.noaa.gov/emc/pages/numerical_forecast_systems/hrrr.php.

84. "Part 3—CBAM, When Models Go Bad Ass," Tomorrow.io, May 8, 2019, https://www.tomorrow.io/blog/cbam-launch/.

85. This exact wording comes from the National Weather Service website (https://www.weather.gov/about/), though many meteorologists speak casually about saving lives and protecting property as focus and motivation. It's a touchstone similar to the unofficial "Neither snow nor rain..." credo of the US Postal Service.

86. "GFS Forecast Model: Surface Temperature—Real-Time," National Oceanic and Atmospheric Administration, February 25, 2016, https://sos.noaa.gov/catalog/datasets/gfs-forecast-model-surface-temperature-real-time/.

87. "National Digital Forecast Database Definitions," National Oceanic and Atmospheric Administration, accessed May 5, 2024, https://digital.weather.gov/staticpages/definitions.php.

88. "Northwest Sustainable Agroecosystems Research," US Department of Agriculture, June 26, 2018, https://www.ars.usda.gov/pacific-west-area/pullman-wa/northwest-sustainable-agroecosystems-research/docs/location/.

89. You can find overview information from Microsoft at https://www.microsoft.com/en-us/garage/wall-of-fame/farmbeats/. The company has since incorporated it into its Azure cloud-computing offerings (https://azure.microsoft.com/en-us/products/data-manager-for-agriculture/). A more academic treatment can be found here: Deepak Vasisht et al., "FarmBeats: An IoT Platform for Data-Driven Agriculture," paper presented at 2017 USENIX Symposium on Networked Systems Design and Implementation (NSDI '17), Boston, Massachusetts, March 2017, https://www.microsoft.com/en-us/research/publication/farmbeats-iot-platform-data-driven-agriculture/.

90. According to the Environmental Protection Agency, in 2021 agriculture accounted for 10 percent of US greenhouse-gas emissions, while on a worldwide basis, that number is generally considered higher. Methane produced by cattle is relatively well known, but, the EPA says, "management of agricultural soils accounts for just over half of the greenhouse gas emissions" from agriculture, as increased levels of nitrogen in soil lead to nitrous oxide emissions.

91. For an example of how farmers can play a role in carbon sequestration, consider the practice of no-till farming. Tilling a field creates better conditions for crops to grow, but the process can disturb the soil and release carbon into the atmosphere. No-till

techniques require farmers to work with greater precision on weather-related decisions about when to plant and fertilize.

92. The Federal Communications Commission has more information on TV white space and its possible uses; see https://www.fcc.gov/general/white-space.

93. Peeyush Kumar et al., "Micro-Climate Prediction-Multi Scale Encoder-Decoder Based Deep Learning Framework," paper presented at 2021 ACM SIGKDD Conference on Knowledge Discovery and Data Mining, Singapore, August 2021, https://doi.org/10.1145/3447548.3467173.

94. More technical details can be found here: "Comparison of the Weather Forecast: Scientific Model of Ignitia and Competitors," May 2023, https://ignitia.se/wp-content/uploads/2023/05/Comparison-of-the-weather-forecast-ENG_.pdf.

95. Details on the approach can be found at https://tempest.earth/tempest-forecasting/.

96. "SETI@home Going into Hibernation," SETI Institute, March 4, 2020, https://www.seti.org/setihome-going-hibernation.

97. See, for instance, Rosetta@home and DENIS@home.

98. As someone old enough to remember car trips with paper maps, I'm fascinated by the reliability of the complex GPS technology. For more on how it works, see "Satellite Navigation—GPS—How It Works," Federal Aviation Administration, June 24, 2022, https://www.faa.gov/about/office_org/headquarters_offices/ato/service_units/techops/navservices/gnss/gps/howitworks.

99. Geodesy, a branch of applied mathematics, focuses on the shape and size of the Earth and the position of points on its surface, so GNSS data helps power the work of scientists like Soja.

100. Benedikt Soja et al., "Machine Learning–Based Exploitation of Crowdsourced GNSS Data for Atmospheric Studies," paper presented at 2023 IEEE International Geoscience and Remote Sensing Symposium, Pasadena, California, July 2023, https://doi.org/10.1109/IGARSS52108.2023.10283441.

101. "What's in the Forecast: Using Cutting-Edge Weather Research to Advance the Waymo Driver," Waymo, November 14, 2022, https://waymo.com/blog/2022/11/using-cutting-edge-weather-research-to-advance-the-waymo-driver.

102. Adam Grossman, "How Dark Sky Works," *Jackadamblog*, November 7, 2011, https://jackadam.github.io/2011/how-dark-sky-works/.

103. The terminology underlying the acronyms varies around the world. In the United States, a METAR is short for "meteorological aviation report," but it is sometimes referred to as an "aviation routine weather report." TAF is a "terminal area forecast," though it's sometimes referred to as a "terminal aerodrome forecast." The specific terms that form the acronyms don't really matter, as they're always referred to as METARs and TAFs.

5. EXTREME HEAT

104. For more on the history and significance of Shockoe Bottom and its role in the slave trade, a good resource is the website of the Shockoe Project (https://theshockoeproject.com), a recent effort, backed by Richmond's city government, to preserve and contextualize the site. For an overview of the records at the Library of Virginia and for information on how to search digitized records, visit Virginia Untold (https://lva-virginia.libguides.com/virginia-untold).

105. There are also non-airport locations. For instance, not far from my home in New York City, there's an ASOS in Central Park at Belvedere Castle, a landmark decorative structure atop a high rock outcrop. The Weather Service has been recording conditions there since 1918; a modern ASOS was installed in 1995. More information on the ASOS network can be found at https://www.ncei.noaa.gov/products/land-based-station/automated-surface-weather-observing-systems.

106. It's interesting to note the corresponding property-damage numbers. Heat was blamed for $6 million in damages, while tornadoes did $2.5 billion in damage—more than four hundred times more. See "Summary of Natural Hazard Statistics for 2020 in the United States," National Weather Service, April 8, 2022, https://www.weather.gov/media/hazstat/sum20.pdf.

107. For a discussion of this, see *Climate Change 2023: Synthesis Report. Contribution of Working Groups I, II and III to the Sixth Assessment Report of the Intergovernmental Panel on Climate Change*, United Nations Intergovernmental Panel on Climate Change, March 20, 2023, https://www.ipcc.ch/report/ar6/syr/downloads/report/IPCC_AR6_SYR_FullVolume.pdf.

108. "Residential Energy Consumption Survey," US Energy Information Administration, https://www.eia.gov/consumption/residential/.

109. *Pavement Irrigation to Improve Thermal Comfort: Métropole Nice Côte d'Azur, France*, EU Covenant of Mayors for Climate and Energy, June 2020, https://eu-mayors.ec.europa.eu/sites/default/files/2022-10/eumayors-case study-Nice-en-2020.pdf.

110. Tree cover figures and other interesting facts can be found in the city Hunting Park plan: *Beat the Heat Hunting Park: A Community Heat Relief Plan*, City of Philadelphia Office of Sustainability, July 19, 2019, https://www.phila.gov/media/20190719092954/HP_R8print-1.pdf.

111. Ziad Obermeyer, Jasmeet K. Samra, and Sendhil Mullainathan, "Individual Differences in Normal Body Temperature: Longitudinal Big Data Analysis of Patient Records," *BMJ* 359 (December 2017), https://doi.org/10.1136/bmj.j5468. Also see Myroslava Protsiv et al., "Decreasing Human Body Temperature in the United States Since the Industrial Revolution," *eLife* 9 (January 2020), https://doi.org/10.7554/eLife.49555.

112. Useful information on symptoms of heat illnesses and when to seek medical help can be found on many consumer-focused online resources, such as WebMD. For a more

detailed overview, see Robert Gauer and Bryce K. Meyers, "Heat-Related Illnesses," *American Family Physician* 99, no. 8 (April 2019): 482–89.

113. You might recall from high-school physics that heat reflects the movement of particles in matter. Heat an object, and the added energy will cause particles to move more quickly. Cooling the object—removing energy—results in particles slowing down. In the case of a gas, heating the particles, causing them to move more energetically, will increase the pressure in a given volume.

114. *Requirements and Standards for NWS Climate Observations*, National Weather Service, November 9, 2023, https://www.nws.noaa.gov/directives/sym/pd01013002curr.pdf.

115. R. G. Steadman, "The Assessment of Sultriness. Part I: A Temperature-Humidity Index Based on Human Physiology and Clothing Science," *Journal of Applied Meteorology and Climatology* 18, no. 7 (July 1979): 861–73, https://doi.org/10.1175/1520-0450(1979)018<0861:TAOSPI>2.0.CO;2.

116. "Heat Index Chart," National Weather Service, May 2022, https://www.noaa.gov/sites/default/files/2022-05/heatindex_chart_rh.pdf.

117. Despite the seeming simplicity of the chart, Steadman's underlying work is highly complex, factoring in everything from the surface area of a typical human body to how clothing affects heat transfer. In a July 1990 National Weather Service technical memo, Lans P. Rothfusz wrote that with summer, "NWS phones are ringing off their hooks with questions about the Heat Index [. . .] Some callers are satisfied with the response that it is extremely complicated."

118. "Heat Hazard Recognition," Occupational Safety and Health Administration, accessed April 18, 2024, https://www.osha.gov/heat-exposure/hazards#environmentalheat.

119. For a guide to the heat index as well as the criteria for heat-related warnings, see "Heat Forecast Tools," National Weather Service, accessed April 18, 2024, https://www.weather.gov/safety/heat-index.

120. "NWS HeatRisk Prototype," National Weather Service, accessed April 18, 2024, https://www.wrh.noaa.gov/wrh/heatrisk/.

121. Though anyone can access WPC forecasts online, they're typically relied on by weather insiders, from Weather Service forecasters in local offices to local officials and emergency managers. For more context, see the WPC's future plans: "The Weather Prediction Center: Five Year Roadmap (2022–2027)," National Weather Service, accessed April 18, 2024, https://www.wpc.ncep.noaa.gov/WPC_Roadmap.pdf.

122. "GCCHE Climate and Health Rapid Response—Heatwaves in the Pacific Northwest," Global Consortium on Climate and Health Education, August 26, 2021, YouTube, https://www.youtube.com/watch?v=vXMkz2Hnpa8.

123. Data on air-conditioning can be found in the US Census American Housing Survey, available at https://www.census.gov/programs-surveys/ahs.html.

124. Their rapid-response analysis, released only days later, can be found at "Western North American Extreme Heat Virtually Impossible Without Human-Caused Climate

Change," World Weather Attribution, July 7, 2021, https://www.worldweatherattribution.org/western-north-american-extreme-heat-virtually-impossible-without-human-caused-climate-change/. In 2022 scientists from the project published the analysis in the peer-reviewed journal *Earth System Dynamics*.

125. Sadly, van Oldenborgh died just a few months after the 2021 heat dome, following a yearslong battle with cancer.

126. "Heat Wave 2021," Washington State Department of Health, accessed April 18, 2024, https://doh.wa.gov/emergencies/be-prepared-be-safe/severe-weather-and-natural-disasters/hot-weather-safety/heat-wave-2021.

127. *Initial After-Action Review of the June 2021 Excessive Heat Event*, State of Oregon Office of Emergency Management, July 27, 2021, https://www.oregon.gov/oem/Documents/2021_June_Excessive_Heat_Event_AAR.pdf.

128. *Extreme Heat and Human Mortality: A Review of Heat-Related Deaths in B.C. in Summer 2021*, British Columbia Coroners Service, June 7, 2022, https://www2.gov.bc.ca/assets/download/9224523E92B045A49942F17F1D40EFEA.

129. Jackson Voelkel et al., "Assessing Vulnerability to Urban Heat: A Study of Disproportionate Heat Exposure and Access to Refuge by Socio-Demographic Status in Portland, Oregon," *International Journal of Environmental Research and Public Health* 15, no. 4 (April 2018): 640, https://doi.org/10.3390/ijerph15040640.

130. If you're curious to see whether there's information on your own community, the campaigns and their reports can be found at https://www.heat.gov/pages/nihhis-urban-heat-island-mapping-campaign-cities.

131. A helpful history that dissects the process used to create the redlining maps can be found in this report by the National Community Reinvestment Coalition, a nonprofit that advocates for fair lending practices: Bruce Mitchell and Juan Franco, *HOLC "Redlining" Maps: The Persistent Structure of Segregation and Economic Inequality*, National Community Reinvestment Coalition, February 2018, https://ncrc.org/wp-content/uploads/dlm_uploads/2018/02/NCRC-Research-HOLC-10.pdf.

132. Jeremy S. Hoffman, Vivek Shandas, and Nicholas Pendleton, "The Effects of Historical Housing Policies on Resident Exposure to Intra-Urban Heat: A Study of 108 US Urban Areas," *Climate* 8, no. 1 (January 2020): 12, https://doi.org/10.3390/cli8010012.

133. Joan Ballester et al., "Heat-Related Mortality in Europe During the Summer of 2022," *Nature Medicine* 29, no. 7 (July 2023): 1857–66, https://doi.org/10.1038/s41591-023-02419-z.

134. "Annual WPC Mean Absolute Errors: Maximum Temperatures," Weather Prediction Center, accessed April 19, 2024, https://www.wpc.ncep.noaa.gov/images/hpcvrf/maemaxyr.gif.

135. *National Emphasis Program—Outdoor and Indoor Heat-Related Hazards*, Occupational Health and Safety Administration, April 8, 2022, https://www.osha.gov/sites/default/files/enforcement/directives/CPL_03-00-024.pdf.

136. "King County to Develop Its First-Ever Extreme Heat Mitigation Strategy to Prepare the Region for More Intense, Prolonged Heat Waves Caused by Climate Change," King County, Washington, June 24, 2022, https://www.kingcounty.gov/en/legacy/depts/dnrp/newsroom/newsreleases/2022/june/24-extreme-heat-mitigation-strategy.

137. For a summary of the Richmond heat-mapping results from 2021, see *Virginia Foundation for Independent Colleges Heat Watch Report*, CAPA Strategies, March 2022, https://www.vfic.org/wp-content/uploads/2022/03/Heat-Watch-VFIC_Report_110321.pdf. On this interactive map, you can zoom in anywhere to find granular temperature data: https://www.arcgis.com/apps/webappviewer/index.html?id=434520e783934ced90b7efbda74f1328.

6. HURRICANES

138. The damage estimate is from "NOAA National Centers for Environmental Information (NCEI) U.S. Billion-Dollar Weather and Climate Disasters." The fatality number is from *Tropical Cyclone Report: Hurricane Ian*, National Hurricane Center, April 3, 2023, https://www.nhc.noaa.gov/data/tcr/AL092022_Ian.pdf.

139. Kevin A. Reed, Michael F. Wehner, and Colin M. Zarzycki, "Attribution of 2020 Hurricane Season Extreme Rainfall to Human-Induced Climate Change," *Nature Communications* 13, no. 1 (April 12, 2022): 1905, https://doi.org/10.1038/s41467-022-29379-1.

140. "Global Warming and Hurricanes: An Overview of Current Research Results," NOAA Geophysical Fluid Dynamics Laboratory, April 17, 2024, https://www.gfdl.noaa.gov/global-warming-and-hurricanes/. See also Knutson et al., "Tropical Cyclones and Climate Change Assessment: Part II."

141. Thomas R. Knutson et al., "Global Projections of Intense Tropical Cyclone Activity for the Late Twenty-First Century from Dynamical Downscaling of CMIP5/RCP4.5 Scenarios," *Journal of Climate* 28, no. 18 (September 15, 2015): 7203–24, https://doi.org/10.1175/JCLI-D-15-0129.1. See also Michael F. Wehner et al., "Changes in Tropical Cyclones Under Stabilized 1.5 and 2.0°C Global Warming Scenarios as Simulated by the Community Atmospheric Model under the HAPPI Protocols," *Earth System Dynamics* 9, no. 1 (February 28, 2018): 187–95, https://doi.org/10.5194/esd-9-187-2018.

142. The death toll has been revised over the years. As of 2023, the National Hurricane Center's figure stands at 1,392 deaths—including direct and indirect deaths, as well as those from indeterminate causes. *Tropical Cyclone Report: Hurricane Katrina*, National Hurricane Center, January 4, 2023, https://www.nhc.noaa.gov/data/tcr/AL122005_Katrina.pdf.

143. Ahrens and Henson, *Meteorology Today*.

144. I've simplified here, since the actual figure depends on aircraft weight, flaps configuration, and other variables, and I've converted from knots to miles per hour. But hopefully this still serves as a point of reference for the ferocity of hurricane winds.

145. Sebastiaan N. Jonkman et al., "Brief Communication: Loss of Life Due to Hurricane

Harvey," *Natural Hazards and Earth System Sciences* 18, no. 4 (April 19, 2018): 1073–78, https://doi.org/10.5194/nhess-18-1073-2018.

146. "Hurricane Hugo: An Eyewitness Account," National Weather Service, accessed May 22, 2024, https://www.aoml.noaa.gov/hrd/flyers/StormSurgeSurvival.pdf.

147. Later it was discovered that the high school, which had been believed to be about twenty feet above sea level, was only about ten, making it a poor location to shelter.

148. Nishant Kishore et al., "Mortality in Puerto Rico after Hurricane Maria," *New England Journal of Medicine* 379, no. 2 (July 12, 2018): 162–70, https://doi.org/10.1056/NEJMsa1803972.

149. The same storm was also known as Super Typhoon Yolanda. The two names reflect different meteorological groups; the Japan Meteorological Agency, which watches over the Pacific, chose Haiyan, while the Philippines' weather officials went with Yolanda.

150. *Tropical Cyclone Report: Hurricane Irma,* National Hurricane Center, September 24, 2021, https://www.nhc.noaa.gov/data/tcr/AL112017_Irma.pdf.

151. "Service Assessment: Hurricane/Post-Tropical Cyclone Sandy, October 22–29, 2012," National Weather Service, https://www.weather.gov/media/publications/assessments/Sandy13.pdf.

152. *Tropical Cyclone Report: Hurricane Sandy,* National Hurricane Center, February 12, 2013, https://www.nhc.noaa.gov/data/tcr/AL182012_Sandy.pdf.

153. Clifford Mass, "The Uncoordinated Giant: Why U.S. Weather Research and Prediction Are Not Achieving Their Potential," *Bulletin of the American Meteorological Society* 87, no. 5 (May 2006): 573–84, https://doi.org/10.1175/BAMS-87-5-573.

154. Louis W. Uccellini et al., "EPIC as a Catalyst for NOAA's Future Earth Prediction System," *Bulletin of the American Meteorological Society* 103, no. 10 (October 2022): e2246–64, https://doi.org/10.1175/BAMS-D-21-0061.1.

155. The nominee, Barry Myers, was a controversial figure due to his having been CEO at AccuWeather, which had a history of pushing for limits on National Weather Service forecasts that would benefit private forecasters. Biden nominated Rick Spinrad, a longtime NOAA scientist, who was quickly confirmed.

156. The core idea—taking advantage of the thousands of commercial flights each day to obtain weather data for models normally acquired with balloons—proved successful. During the COVID-19 pandemic, curtailed air travel translated to a reduction in weather data, and this is why.

157. Six months later, Americans would watch a much larger version of this discord play out when the COVID-19 pandemic became politicized, leading to distrust of both vaccines and scientifically supported medical advice.

158. Peter Baker, Lisa Friedman, and Christopher Flavelle, "Trump Pressed Top Aide to Have Weather Service 'Clarify' Forecast That Contradicted Trump," *New York Times*, September 11, 2019, https://www.nytimes.com/2019/09/11/us/politics/trump-alabama-noaa.html.

159. Andrew Freedman et al., "Trump Pushed Staff to Deal with NOAA Tweet That Contradicted His Inaccurate Alabama Hurricane Claim, Officials Say," *Washington Post*, November 8, 2019, https://www.washingtonpost.com/weather/2019/09/11/lawmakers-commerce-department-launch-investigations-into-noaas-decision-back-presidents-trump-over-forecasters/.

160. "Evaluation of NOAA's September 6, 2019, Statement About Hurricane Dorian Forecasts," Office of the Inspector General, June 26, 2020, https://www.oig.doc.gov/OIGPublications/OIG-20-032-I.pdf.

161. Edmund P. Willis and William H. Hooke, "Cleveland Abbe and American Meteorology, 1871–1901," *Bulletin of the American Meteorological Society* 87, no. 3 (March 2006): 315–26, https://doi.org/10.1175/BAMS-87-3-315.

162. Bjerknes's landmark 1904 paper "Das Problem der Wettervorhersage, betrachtet vom Standpunkte der Mechanik und der Physik" ("The Problem of Weather Prediction, Considered from the Viewpoints of Mechanics and Physics") is prescient not only for its conception of numerical prediction but also for its appreciation that new observations could help drive this work. "By means of radiotelegraphy, it will be possible to include among the reporting stations steamships with fixed routes," Bjerknes wrote, according to this 2009 translation by Esther Volken and Stefan Brönnimann; see Vilhelm Bjerknes, "The Problem of Weather Prediction, Considered from the Viewpoints of Mechanics and Physics," *Meteorologische Zeitschrift* 18 (6): 663–67, https://doi.org/10.1127/0941-2948/2009/416.

163. Richardson toiled over what are known as Navier-Stokes equations, which describe fluid motion and underlie modern numerical prediction.

164. After it left our planet, GOES-U underwent a name change. The satellites are designated by a letter prior to launch but switch to a number once in space. GOES-U was launched on June 25, 2024. It is now known as GOES-19.

165. As of this writing, NASA has slated 2032 for the launch of the first of those successor craft.

166. A problem with GOES-S, which launched in 2018, illustrated the approach. A flaw in the cooling system for its key sensor, the Advanced Baseline Imager, wasn't discovered until the satellite had reached orbit. This glitch affected the ABI's ability to measure infrared radiation, an important indicator of energy in the atmosphere and a key tool for understanding conditions at night. Though most of the instrumentation worked fine, and the cooling system glitch affected only the infrared data during some overnight hours at certain times of the year, the problem required NASA to rely on its previous-generation satellite for some observations of the western United States. Engineers at L3Harris Technologies, which built the ABI, redesigned the cooling system for GOES-T and GOES-U. Eventually GOES-T took over the GOES-West duties and GOES-S moved to a standby role.

167. John Cangialosi, *National Hurricane Center Forecast Verification Report: 2022 Hurricane Season*, National Hurricane Center, May 8, 2023, https://www.nhc.noaa.gov/verification/pdfs/Verification_2022.pdf.

168. *Tropical Cyclone Report: Hurricane Laura*, National Hurricane Center, May 26, 2021, https://www.nhc.noaa.gov/data/tcr/AL132020_Laura.pdf.

169. "A Tornado Hits the Weather Channel," YouTube, https://www.youtube.com/watch?v=0cODBQqaGTw.

170. At one point the Weather Channel was owned by a group that included NBC Universal. In the internet era, it created Weather.com, a go-to digital source for forecasts and news. In 2015, IBM agreed to purchase the parent entity, the Weather Company, but left the TV network out of the deal; the tech giant wanted the digital assets as part of its artificial intelligence future but didn't want to run a television operation. (Weather.com continued to brand itself "the Weather Channel" under a licensing arrangement; IBM has since sold the Weather Company to a private-equity firm.) In 2018, the privately held Allen Media Group agreed to buy the Weather Channel for a reported $300 million.

171. Joseph E. Trujillo-Falcón et al., "¿Aviso o Alerta? Developing Effective, Inclusive, and Consistent Watch and Warning Translations for U.S. Spanish Speakers," *Bulletin of the American Meteorological Society* 103, no. 12 (December 2022): e2791–2803, https://doi.org/10.1175/BAMS-D-22-0050.1.

172. Richard Van Noorden and Jeffrey M. Perkel, "AI and Science: What 1,600 Researchers Think," *Nature* 621, no. 7980 (September 28, 2023): 672–75, https://doi.org/10.1038/d41586-023-02980-0.

173. In academic publishing, a preprint refers to an early-release version of an article that hasn't yet completed the peer-review process necessary to appear in most scientific journals. Preprints allow researchers to get their results out quickly, giving other scientists an opportunity to look at the work in a fast-moving field and also helping to establish bragging rights in terms of who accomplished what when.

174. Kaifeng Bi et al., "Accurate Medium-Range Global Weather Forecasting with 3D Neural Networks," *Nature* 619, no. 7970 (July 20, 2023): 533–38, https://doi.org/10.1038/s41586-023-06185-3.

175. Remi Lam et al., "Learning Skillful Medium-Range Global Weather Forecasting," *Science* 382, no. 6677 (December 22, 2023): 1416–21, https://doi.org/10.1126/science.adi2336.

176. Computation energy usage and its environmental effects have gotten more attention when it comes to cryptocurrencies, which are "mined" by complicated calculations. A February 2024 statement from the US Energy Information Administration estimated that the massive crypto industry may account for 0.6 to 2.3 percent of all annual US energy consumption.

7. SEASONAL FORECASTING

177. "U.S. to Feed Half a Million Zimbabweans Until October After Drought," Reuters, April 20, 2016.

178. In addition to decimating crops, the drought also contributed to shortages of drinking water; see "Malnutrition on the Rise in Zimbabwe as El Niño Takes Its Toll," UNICEF

USA, March 15, 2016, https://www.unicefusa.org/press/malnutrition-rise-zimbabwe-el-nino-takes-its-toll.

179. "United States Provides Additional $54.5 Million for Humanitarian Assistance in Zimbabwe," US Agency for International Development, August 4, 2016, https://2012-2017.usaid.gov/zimbabwe/press-releases/united-states-provides-additional-545-million-humanitarian.

180. You can see these for yourself at https://www.cpc.ncep.noaa.gov/products/predictions/WK34/.

181. "How the *Old Farmer's Almanac* Predicts the Weather," *Old Farmer's Almanac*, November 22, 2023, https://www.almanac.com/how-old-farmers-almanac-predicts-weather.

182. "What Is Global Circulation? Part Three: The Coriolis Effect & Winds," UK Met Office, March 16, 2018, YouTube video, https://www.youtube.com/watch?v=PDEcAxfSYaI.

183. Herbert Riehl and Dave Fultz, "Jet Stream and Long Waves in a Steady Rotating-Dishpan Experiment: Structure of the Circulation," *Quarterly Journal of the Royal Meteorological Society* 83, no. 356 (April 1957): 215–31, https://doi.org/10.1002/qj.49708335608.

184. This origin story of the term *El Niño* is widely accepted in meteorology circles. For this and the basics of how the effect occurs and fits into the large ENSO pattern, I'm indebted again to Ahrens and Henson's excellent introductory college textbook *Meteorology Today*. You can find helpful overviews online from NASA, NOAA, and the UK Met Office.

185. The Madden-Julian Oscillation takes its name from Roland Madden and Paul Julian, the scientists at the National Center for Atmospheric Research credited with discovering the phenomenon and describing it in a 1971 paper.

186. Christopher J. White et al., "Advances in the Application and Utility of Subseasonal-to-Seasonal Predictions," *Bulletin of the American Meteorological Society* 103, no. 6 (June 2022): e1448–72, https://doi.org/10.1175/BAMS-D-20-0224.1.

187. I've relied on interviews with program leaders and participants to describe the drought response in Zimbabwe, but for those interested in a broader overview of anticipatory-action efforts across the world, it's worth reading this report: *Scaling Up Anticipatory Actions for Food Security: Anticipatory Action Year in Focus 2023*, World Food Programme, April 2024, https://www.wfp.org/publications/scaling-anticipatory-actions-food-security-anticipatory-action-year-focus-2023.

188. *Economics of Resilience to Drought: Kenya Analysis*, US Agency for International Development, January 4, 2018, https://2017-2020.usaid.gov/sites/default/files/documents/1867/Kenya_Economics_of_Resilience_Final_Jan_4_2018_-_BRANDED.pdf.

189. *Forecast-Based Financing in Nepal: A Return on Investment Study*, World Food Programme, May 2019, https://www.wfp.org/publications/forecast-based-financing-nepal-return-investment-study.

190. *El Niño Anticipatory Action Plan: Zimbabwe Oct. 2023–Mar. 2024*, United Nations Office for the Coordination of Humanitarian Affairs, November 2023, https://www

.unocha.org/publications/report/zimbabwe/zimbabwe-el-nino-anticipatory-action-plan-oct-2023-mar-2024-issued-november-2023.

CONCLUSION

191. *Tropical Cyclone Report: Hurricane Ida*, National Hurricane Center, April 4, 2022, https://www.nhc.noaa.gov/data/tcr/AL092021_Ida.pdf.

192. In addition to the thirteen direct fatalities, New York City recorded one additional death from indirect causes. Ariel Yuan et al., "Immediate Injury Deaths Related to the Remnants from Hurricane Ida in New York City, September 1–2, 2021," *Disaster Medicine and Public Health Preparedness* 18 (2024): e55, https://doi.org/10.1017/dmp.2024.49.

193. "Transcript: Mayor de Blasio Delivers Remarks with Governor Hochul to Provide Update on New York City," City of New York, September 2, 2021, http://www.nyc.gov/office-of-the-mayor/news/596-21/transcript-mayor-de-blasio-delivers-remarks-governor-hochul-provide-on-new-york.

194. "Is New York City Ready for Rain? An Investigation into the City's Flash Flood Preparedness," Office of the New York City Comptroller, April 2024, https://comptroller.nyc.gov/wp-content/uploads/documents/Is-New-York-City-Ready-for-Rain.pdf.

195. "Climate Change and the Escalation of Global Extreme Heat: Assessing and Addressing the Risks," Climate Central, Red Cross Red Crescent Climate Centre, and World Weather Attribution, May 28, 2024, https://www.climatecentral.org/report/climate-change-and-the-escalation-of-global-extreme-heat.

196. A. R. Crimmins et al., eds., *Fifth National Climate Assessment*, US Global Change Research Program, 2023, https://doi.org/10.7930/NCA5.2023.CH1.

197. For these budget numbers, I have relied on the Federal Science Budget Tracker website from the American Institute of Physics, which summarizes year-by-year data on budget requests and amounts approved; see https://ww2.aip.org/fyi/fy2025-noaa-budget-and-appropriations.

198. CPI-adjusted as of June 2024. See "2023: A Historic Year of U.S. Billion-Dollar Weather and Climate Disasters," National Oceanic and Atmospheric Administration, January 8, 2024, http://www.climate.gov/news-features/blogs/beyond-data/2023-historic-year-us-billion-dollar-weather-and-climate-disasters.

199. Alex Hern, "Twitter Usage in US 'Fallen by a Fifth' since Elon Musk's Takeover," *Guardian*, March 26, 2024, https://www.theguardian.com/technology/2024/mar/26/twitter-usage-in-us-fallen-by-a-fifth-since-elon-musks-takeover.

200. Shawn Hubler, "When the Storm Online Is Worse Than the One Outside," *New York Times*, February 1, 2024, https://www.nytimes.com/2024/02/01/us/storm-misinformation-viral-social.html.

Index

Abbe, Cleveland, 184
Abrams, Stephanie, 199–202
academic world, 92
accuracy of forecasts, measures of, 175
AccuWeather, 6, 77, 90–93, 104, 123, 201
Adams, Eric, 232
Adrienne Arsht-Rockefeller Foundation Resilience Center, 162
Advanced Baseline Imager (ABI), 25–26
Advanced Weather Interactive Processing System (AWIPS), 78–79
aerosols (in the atmosphere), 223
Africa, 169, 227
Agee, Ernest, 20
Agelasto, Parker, 129
agriculture, 110, 210–11, 224
Agyemang, Miranda Osei, 113
Ahrens, C. Donald, 16, 170
AirBeam, 130
air-conditioning, 133–34, 153–54, 160
AirDat, 181
air-sea interaction, 217
Alabama, 2, 20, 37–38, 182
alert, 84
Amanor-Larbi, Ernest, 113
Ambient Weather, 117
American Meteorological Society (AMS), 4, 173
Amish population, 82–84, 236
anticipatory action, 212, 224–25

Apple, 109, 123
Arctic Oscillation (AO), 220
area forecast discussion (AFD), 73–74
Argentina, 19
Aristotle, 3
Arizona, 145
Arkansas, 20
artificial intelligence (AI), 3, 59, 109, 203–4, 233, 237
Artificial Intelligence/Integrated Forecasting System (AIFS), 209
atmosphere
    future state of, 4
    sounding of, 46
atomic bombs, 17–18
attribution science, 153–54
automated phone service, 82–83
Automated Surface Observing Systems (ASOs), 97, 130
aviation industry, 18
Avista utility company, 151
Awareness for a Rural Nation (WARN), 83
Aztecs, 215
Azure cloud-computing platform, 109

Banghoff, John, 73, 74–75, 78–79, 82–83, 95–100
Bassill, Nick, 229
Behavioral Insights Unit, 34
BeiDou, 120

INDEX

Berchoff, Don, 126
"better, cheaper, and faster", 102–3
Biden, Joe, administration, 160, 234
Bjerknes, Vilhelm, 184
Blacks, 236
blackouts, 61
blocking event, 150
body bag coolers, 152
body heat, regulation of, 141
Bojorquez, Manuel, 198
Boston University, Center for Climate and Health at, 141
Braun, Peter, 129
Briefing (WFO), 96
British Columbia, Canada, 155
Brown, Alton, 181
Brown, Dan, 194–96
Burke, Patrick, 42
Busalacchi, Antonio, 92
businesses, information needed by, 8
Byers, Horace, 18

Cal Fire, 65–66
Calhoun, Kristin, 27
California, 2
  Department of Forestry and Fire Protection (Cal Fire), 52, 65
call center, 71
CAMALIOT, 120–22
Cameron Peak Fire (2020), 53
Camp Fire of 2018, 52–56
  fully contained, 56
  investigation, 56–57
  lives lost, 56
Canada, 2, 19, 144, 155
Cantore, Jim, 198
CAPA Strategies, 156
Capital Region Land Conservancy, 129
carbon dioxide, in the atmosphere, 2–3
carbon emissions, 221
Caribbean, 194
Carver, George Washington, 28
Caterpillar, 95
Cedar Fire of 2003, 49
Centers for Disease Control (CDC), 146, 156, 199
Central America, 194
certainties, 212–13
Chandra, Ranveer, 107–9
Chaney, Ray, 70–71

Chantry, Matthew, 204–5
chaparral, 48–49
Chapman, Saleem, 135
chaser vehicle, 46
ChatGPT, 205
Cheung, Alex Alvin, 88
children, affected by heat, 160
China, 3, 120, 203, 215
C hook, 54–55
Christian, Hugh, 26–27
chronic health patients, 160
clean room, 185–86
ClimaCell, 113
climate alarmism, 235
climate change, 2, 50, 52, 68, 107–8, 153–54, 166, 190, 234
Climate Prediction Center (CPC), 213–14
climatology, 215–16
clouds, measuring, 98
coastal ecosystems, 49
College of Atmospheric Sciences, 21
College Park, MD, 147–48
Commerce Department, 15, 105
communicating with the public, 196
communities, weather-related decisions of, 6
community organizations
  As I Plant This Seed, 139
  Esperanza, 137
companies, weather-focused, 6
Comprehensive Bespoke Atmospheric Model (CBAM), 104, 114
computer graphics, 198
computers, 66, 69, 164–165, 179, 184, 237
conduction, 141
Conservation Camps Program, 65
convection, 112, 141
convection-allowing model (CAM), 42
cooling centers, 133–34, 160, 162
Cooperative Institute for Severe and High-Impact Weather Research and Operations (CIWRO), 22
Copernicus Climate Change Service, 190
Coriolis, Gaspard-Gustave de, 169
Coriolis effect, 169, 216
Cortes, Luis Jr., 137
Covenant of Mayors for Climate and Energy, 134
COVID-19 pandemic, 134, 137
cowpeas, 224

INDEX

Cray supercomputers, 164–65
crowdsourcing, 120
CubeSats, 115
Cyclone Nargis, (2008), 173
Czapinski, Neil, 67

D'Agostino, Brian, 50, 57–58, 60–61, 71
damage, assessment of, 19
Damush, Jon, 127
Dark Sky, 109, 123
data collection sources, 114
Davis Instruments, 117
deaths, weather-related, 14, 131, 151
de Blasio, Bill, 232
DeepMC, 109
demographic maps, 156
Department of Agricultural, Technical and Extension Services (Agritex), 224
Department of Defense, 125
Destination Earth, 190
deterministic forecasts, short-term, 213
Deutscher Wetterdienst (DWD), Germany, 168
developing nations, 237
development, unchecked, 68
Disaster Relief Appropriations Act of 2013, 178
dishpan experiments, 216–17
Dixie Fire of 2021, 57
Donahoe, Hunter, 88
Doppler radar, 23
Dot weather station, 123
drones, 101–2, 123
DroneUp, 102–3, 126
  delivery mission by, 124
  reports, 125
drought, 211, 226
drownings, 231
duck fountain, 189

Earth,
  rotation of, 216
  satellite view of, 25
Earth Prediction Innovation Center, 180, 182–83
ecosystem, 48–49
Edgington, Samantha, 27
elderly, affected by heat, 160
election of 2024, 8
electric utility, 50–51, 53

Elkabetz, Shimon, 113, 115
Elk City tornado, 42–43
El Niño, 216, 218–19, 227
El Niño Southern Oscillation (ENSO), 218–19
emissions, carbon, 68
energy industry companies, 221
enhanced Fujita scale (EF scale), 19
ensemble forecast, 148, 223
ensembles, 148, 195
Enterprise-Record newspaper, 54
Environmental Modeling Center, 74
equator, 169
ERA5, 209
Estonia, 191
ETEC building, 231
ETH Zurich, 121
"Euro", 188
Euro1k, 123
Euro model, 42
Europe, 19, 159
European Centre for Medium-Range Weather Forecasts (ECMWF), 165, 167–68, 178–79, 182, 188, 190–91, 207, 222, 228
European Space Agency, 120
European Union, 120
European Union-funded Copernicus Climate Change Service, 209
Europoean Center for Medium-Range Weather Forecasts, 42
evacuations, 77, 175, 177
evaporation, 141
Everybody talks about the weather (art exhibition), 10
excessive heat outlook, 145
excessive heat warning, 145
excessive heat watch, 145
extreme heat, 128–63

Facebook, 236
false alarms, 15
fans, 160
FarmBeats, 107–8
farming, 84, 106, 110, 215, 224
FastEddy, 69–70
Federal Aviation Administration (FAA), 23, 125
Finch, Allison, 230–31
finite-volume cubed sphere (FV3), 179

fire, 48–72
    ancient times, 53
    Greeks' view of, 60
fires, extreme, 68
fire-weather watch, 64
five-hundred-millibar (mb) chart, 146
flooding, 199, 226, 228
Florida, 5, 174
Florida Keys, 175
Florida International University, 193
Florida Panhandle, 37–38
Fondazione Prada, 10
Forecast Advisor, 123–24
forecasters, testing of, 25
forecasting, 2, 35, 89
    accuracy of, 175
    future of, 233
    microscale, 70
    nineteenth century, 3
    private, 50, 52
    private and public, 91
    third wave of, 3
    of hurricanes, 4
    of tornadoes, 14
forecasts, 232
    accurate, 197
    acting on, 232–33
    deterministic, 43, 213
    fire-weather, 51
    five-day, 4, 213
    hyperlocal, 104
    one-day, 4
    predictability of, 223
    probabilistic, 7, 43, 213
    real-world, 85
    response to, 3, 6
    seasonal, 213
    subseasonal, 213
forestry agency, 65
FourCastNet, 207
Fox Weather, 201
Fujita, Tetsuya Theodore, 17–19
Fujita scale, 17
Fultz, Dave, 216–17

Gaea supercomputer, 179
Galileo satellite network, 120
Geophysical Fluid Dynamics Laboratory (GFDL) in US, 178–79
Georgia, 2, 20, 37

Geostationary Lightning Mapper (GLM), 26–27
Geostationary Operational Environmental Satellites (GOES), 25, 185
geostationary satellite, 25–26
Ghana, 110
global computer models, 167
Global Consortium on Climate and Health Education, Columbia University, 153
Global Forecast Center, 94
Global Forecast System (GFS), 42, 79, 104, 178, 181
global models, 165–67
global navigation satellite system (GNSS), 121
global NWP models, 204
Global Positioning System (GPS), 120
GLONASS, 120
GOES-R, 25–26, 186–87
GOES-U, 185–86
Google, 205
Gorham, KS, 24
government, services of, 118
government forecasting, 235
GraphCast, 205
graphics processing unites (GPUs), 69
Great Plains, 11, 19–20
Great Zimbabwe University (GZU), 225
greenery, 134
greenhouse effect, 2
greenhouse gas emissions, 161
GRID-Arendal, 68
Grossman, Adam, 123
Guejito Fire of 2007, 58
Gulf of Mexico, 19
gyres, 217–18
GZU Campus Radio, 225

hail, 11, 12
Harborview Medical Center, 152–53
Harris, Ryan, 139–40
Hawaii, 2
hay harvesting, 84
Hazardous Weather Testbed, 25
heat
    and medical conditions, 142
    dangers of, 140
heat advisory, 145

INDEX

heat domes, 150
heat exhaustion, 142
heat index, 81, 144
heat-mapping, 156–57
   campaigns, 156–57
heat maps, 158
heat-respite area, 133
HeatRisk, 146
heatstroke, 142
heat waves, 2
   assigning names to, 162
hemispheres (of earth), 169
Henson, Robert, 16, 170
Hess, Jeremy, 151
Hewson, Tim, 192–93, 203–4
High-Resolution Rapid Refresh model (HRRR), 104, 181
highway construction, 163
Hiroshima, 18
Hochul, Kathy, 229, 233
Hoffman, Jeremy, 157–58
homeless people, 160
Home Owners Loan Corporation (HOLC), 157
Hosler, Charles, 91
House Committee on Science, Space and Technology, 92
Huawei, 203
Huff, C.J., 29
Huhn, Anna Lena, 227–28
humans, processing of probabilistic information, 44
humidex, 144
humidity, 81, 143
   relative and absolute, 144
Hunting Park, Philadelphia, PA, 135–39
hurricane, 171
   eye of, 170
   forecasting, 166–67
   related terms, 170
   structure of a, 170
Hurricane Dorian, 182
Hurricane Harvey, 4, 171, 174
Hurricane Hilary, 7
Hurricane Hugo, 172
Hurricane Hunter, 194
Hurricane Ian, 166
Hurricane Ida, 231–32
Hurricane Irma, 4, 174–75
Hurricane Katrina, 166

Hurricane Laura, 196–97
Hurricane Maria, 4, 5, 172–73, 174
Hurricane Patricia, 171
Hurricane Sandy, 173
hurricanes, 2, 164–209, 165, 169
   massive response to, 63
hurricane season, 165, 168
   2017, 4
hurricane warning, 196
hurricane watch, 196
Hurricane Weather Research and Forecasting Model (HWRF), 194
hydrology, 228
hyperlocal forecasts, 5
hyperlocal weather, 103–4

IBM, 207
ice storm, 2
Ignitia, 110–12
Illinois, 14
immersive mixed reality (IMR), 198
India, 2
Indiana, 14
Indian Ocean, 25
Indian Ocean Dipole (IOD), 220
inequities, 236
Iñiguez, Paul, 145
instrumentation, 12
Integrated Forecasting System (IFS), 188–89, 209, 222
International Research Institute for Climate and Society, 221
internet access, 108, 114
Internet of Things (IoT), 3, 104, 114, 122
iPhone, 109
Iris Automation, 126–27
Israel, 191

Jacobs, Neil, 181–82, 182
Jarbo winds, 55
Jefferson, Thomas, 97
JetBlue, 115
jet stream, 150
Joel N. Myers Weather Center, 90
Johnson, Priscilla, 138–39
Joplin, MO, 33
   tornado EF5 of 2011, 14, 28–29
journalism, as mission, 9
Journal of Applied Meteorology and Climatology, 20

INDEX

Kansas, 20
katabatic winds, 60
Kenney, Jim, 135
Kenya, 226
Kettering, Matt, 186
Keys, FL, 175
King County, WA, 161

Lamers, Alex, 147
Lander, Brad, 232
La Niña, 218–19
Lapenta, William, 181–82
layered-threat, 154
Lee, Amy, 153
Lee County, AL, 39
    disaster, 40
Leininger, Suzann, 64–65
LIDAR technology, 104
lightning, 26–27, 117
lightning jump, 26–27
Lin, Shian-Jiann, 178–79
literacy, 45
Lockheed Martin Space, 185
London, UK, 159
long-range weather forecasts, 98, 211–212
Los Angeles, CA, 161
Louisiana, 20
Lyons, Buck, 118

machine learning for weather, 203–205
Madden-Julian Oscillation (MJO), 220
maize, 224
malaria, 221
Malawi, 2
map apps, 121
Mark Twain, 1
Martin, Lockheed, 27
Mass, Cliff, 180
Maui, HI, 2
Maunganidze, Golden, 225
Mayans, 215
McBride, David, 38
McCarthy, Abigail, 88
McClain, Kim Klockow, 34–36
McGovern, Amy, 206–7
McKenzie, Matt, 55
mesocyclone, 17
mesonet, 13, 104
mesoscale, 13
mesoscale network, 13

mesovortex, 17
METARs, 125
Meteo 415, 86
    forecasting practicum, 85
Meteodrones, 123
Meteomatics, 105, 123
Meteorological, 2
meteorologist, as communicator, 99–100
meteorology, careers in, 87
Meteorology Today (textbook), 16, 170
Met Office, UK, 168
Metts, George, 172
Miami, FL, 135
Miami-Dade County, FL, 161
microgrids, 69
Microsoft, 105, 108–9, 109
microweather, 126
Midland Radio, 84
Mississippi, 20
Missoula Fire Sciences Laboratory
    (Montana), 70
Missouri, 14, 33
mobile homes, 36, 40–41
    shelters for residents of, 44
models, 42, 170, 180–81, 183
    choosing, 193
    consensus models, 195
    global-weather, 70
    in-house forecast, 59
    variety of, 194
Montalto, Franco, 134–36
Moore, OK, 24
Moore, Debra, 151
Morocco, 191
mosquitoes, 221
Mozambique, 2, 228
    National Institute of Meteorology, 228
Mudavanhu, Champion, 225
Mugabe, Robert, 210
Muñoz-Esparza, Domingo, 70
Mupuro, John, 212
Musk, Elon, 115, 236
Myanmar, 173
Myers, Barry, 92–93
Myers, Joel, 90–91, 92

Nagasaki, 18
NASA, 207
National Blend of Models (NBM), 79, 87

INDEX

National Center for Atmospheric Research (NCAR), 69–70, 92
National Centers for Environmental Prediction, College Park, MD, 180
National Climate Assessment, 234
National Hurricane Center, Miami, FL, 4, 193–194
National Integrated Heat Health Information System (NIHHIS), 156
National Lightning Detection Network (NLDN), 28
National Oceanic and Atmospheric Administration (NOAA), 3, 8, 15, 21, 23, 42, 73, 78, 156, 167, 180–81, 207, 234–35
National Severe Storms Laboratory (NSSL), 11–12, 21–22
National Weather Center (NWC), 21
National Weather Service (NWS), 8, 9, 14–15, 24, 27, 30, 71, 74–75, 92, 104, 105, 125, 146, 174, 234–35
   forecasts from, 77–78
   guiding principles, 9
   overall guidelines, 145
   predictions, 155
   proclamations regarding heat, 145
   radar-station network, 124
   service assessment, 30–31
Navarro, Vicente, 120
Ncube, Moffat, 224–25
near term, 98
Nebraska, 20
Nelson, Andrew, 106–9
Nepal, 226
Nese, Jon, 86–87
New Deal, 157
New England, 20
New Orleans, LA, 166
New York, NY, 116
   Central Park, 97
   City, 2, 177
   Sandy, 173
New York state, 230
   Home Energy Assistance Program, 160
   State Weather Risk Communication Center (SWRCC), 229–30, 232
Next Generation Weather Radar (NEXRAD), 23, 45, 124, 235
Nigeria, 221
Norman, OK, 21, 24
Norman weather forecast office (WFO), 22
Norman WFO, 42
North Atlantic Oscillation (NAO), 220
NSF AI Institute for Research on Trustworthy AI in Weather, Climate, and Coastal Oceanography, 206
numerical weather-prediction (NWP) models, 114, 123, 165, 183–84
NVIDIA, 207

Occupational Safety and Health Administration (OSHA), 160
oceans, 217
Okainbea, Mary, 110–11
Oklahoma, 11, 20, 42
Oklahoma City, 14
Oklahoma City twister (1998), 23
Oklahoma Climatological Survey, 22
Oklahoma Mesonet, 22
Old Farmer's Almanac, The, 215
omega block, 150
open-source software development, 181
Ophelia, Tropical Storm, 232
optimism bias, 33
Ortega, Kiel, 12, 45–46
Otto, Friederike, 153–54
outdoor workers, 160

Pacific Decadal Oscillation (PDO), 219
Pacific Gas & Electric Company (PG&E), 54
Pacific Northwest, 7, 20
   heat wave of 2021, 146–55
Palm Sunday tornadoes (1965), 18
Palouse, the, 106–7
Pangu-Weather, 203–4, 207
Pappenberger, Florian, 189
Paradise, CA, 54
pavement wetting, 134
Pendleton, Nicholas, 158
Penn State, 73–76, 80, 85, 87
   Department of Meteorology and Atmospheric Science, 85–86
Pennsylvania Amish Safety Committee, 83
PG&E, 57, 68–69
   investigation, 57
phased-array radar, 23, 235
Philadelpgia More Beautiful Committee, 138
Philadelphia, PA, 135–40, 160
Philippines, 173
photographs from space, 187

Piltz, Steve, 75
planning for weather events, 6, 44–45
   previously agreed-on, 211–12
planter benches, 133, 140
Porter, Jon, 94
Portland, OR, 149, 151
   State University, 156
Portugal, 159
Potts, Michael, 201
power companies, 68
power lines, burying, 68
power shutoffs, 53, 68–69
predictions, 67
   gains in prescience, 5
prison inmates, 65
private sector, 91–92, 93
privatization Project 2025 group, 235
probabilities, 212
product (forecast), 80
public-safety power shutoff (PSPS), 61–62
Puerto Rico, 5, 172–73
Pulga, 55

quality control (QC), 97–98
quasi-linear convective systems (QLCS), 17, 37–39

Rabier, Florence, 188, 207–9
radars, Doppler weather, 124
RadarScope, 45, 124
Radeloff, Volker, 52
radiation, of heat, 141
radios, 83–84
   weather, 45
radiosonde, 12–13
rain, 171
rainfall, 110, 112, 232
Ramirez Cisneros, Joaquin, 66–67
Rasmussen, Erik, 40–41
Raspberry Pi, 117
Reading, U.K., 164, 188–89, 191
readings, temperature (NWS), 143
reanalysis, 209
ReConnect, 108
Red Cross Red Crescent Climate Centre, 228
Redding, CA, 86
redlining, 157–58
renewable energy, 221
resilience officer, 135

response plans, for hospitals and schools, 44–45
Reynhout, Jodi, 137
Richardson, Lewis Fry, 184
Richmond, VA, 128–29, 155, 158, 161, 163
roofs, 161
Ross, Wilbur, 182
Rothfusz, Lans P., 76
Ruiz, Sussy, 202
Russia, 120
RVAgreen 2050, 161
Ryan, Trey, 230

S2S (subseasonal to seasonal), 213, 220–221
Saffir-Simpson Hurricane Wind Scale, 171, 195
   Categories, 172
sagebrush, 49
San Diego, CA, 48, 62, 70–71
San Diego County, 65, 72
San Diego Gas & Electric (SDG&E), 50, 58–59, 71, 72
San Diego Supercomputer Center, 62
Santa Ana Wildfire Threat Index (SAWTI), 62
Santa Ana winds, 50, 58, 60–61
Santorum, Rick, 92–93
satellites, 114–15, 185
   data from, 3
Sauer, Jeremy, 69–70
School of Meteorology, 21
Schultz, Christopher, 27
science, politicization of, 8
Scott, Rick, 5
Search for Extraterrestrial Intelligence (SETI), 119–120
seasonal forecasting, 210–28
Seattle, WA, 149, 151, 152
seawalls, 173
self-driving cars, 123
service assessment, 34
severe weather forecasts, 92
Seville, Spain, 162
shade structures, 134
Shandas, Vivek, 156, 158
Sharpiegate, 182
shifts, 79, 95–96
Short Message Service (SMS), 112
short term, 98
SILVIS Lab, 52

INDEX

simulations, 183
siren sound, 32
skew-T chart, 46
SmallSats, 115
smart apps, 163
smart-home devices, 118
smartphones, 162, 238
social media, 43, 89, 162
   information from, 236
social science, 34
Soja, Benedikt, 121
South Carolina, 37
Southern California, 7, 48
Space Age, 3
space race, 187
SpaceX, 115
spaghetti plots, 193
Spain, 66
Spanish language, 202
Spire Global, 90
Spokane, WA, 151
Spring Experiment, 25
Sputnik, 187
squall lines, 17
Steadman, Robert, 143–44
Stephens, Liz, 228
Stockdale, Tim, 222–23
Stoney, Levar, 161
storm-chasers, 11–13, 124
   amateur, 46–47
Storm Prediction Center (SPC), Norman, OK, 22, 54, 73, 147
Storm Prediction Center (SPC), 22, 147
StormReady, 75–76
storms, 26–27
   naming, 202
storm surge, 172
storm tide, 172
streamline chart, 169
Stroz, Michael, 24, 27–28
Structural Extreme Events Reconnaissance Network (StEER), 40
subseasonal-to-seasonal (S2S) predictive time scale, 220–21
subway tunnel flooding, 177
sunspots, 215
supercell, 17
supercomputers, 179, 184
Superstorm Sandy, 2012, 176–77
surface wind, 106

Sustaining Urban Places Research Lab, 156
swamp coolers, 160
Szatkowski, Gary, 177

TAFs, 125
Tardy, Alex, 62–63, 71
Technosylva, 66–67
teleconnections, 216
telegraph, 3
Tempe, AZ, 162
temperature, 97, 106
   and pressure, 147
   dry-bulb, 144
   measuring, 130
   wet-bulb globe (WBGT), 144–45
Tempest company, 105, 116–17
Tennessee, 20
Texas, 7, 20
Texas Panhandle, 42
text messages, 111
text products, 80–81
The Blue Marble photo, 187
Thomas, Robert B., 215
thumberstorms, 16–17
Thunderstorm Project, 18
TikTok, 236
Tillier, Clem, 27
Time magazine, 24
Tomorrow.io, 90, 104, 105, 113–15
Tornado Alley, 11, 20
   tours of, 47
tornadoes, 2, 11–47
   classification of, 19
   defined, 16
   how it forms, 21
   lead time for warning of, 14–15
   saving more lives from, 41
   training to deal with, 24
tornadogenesis, 21
tornado warning, 29
tornado watch, 29, 39
Totable Tornado Observatory (TOTO), 22–23
track (map), 4
track forecasts, 176
tribal communities, 237
Tri-State Tornado, 14
tropical cyclone, 2, 170
tropical depression, 170
tropical forecasts, 112

tropical storm, 170
  watches and warnings, 196
tropical wave, 169
Trump, Donald, 8, 15, 93, 182, 234
TruWeather Solutions, 126
Turner, Jack, 123
Twister (film), 22–23, 47
Twitter (now X), 236
Two-hundred-and-eleven San Diego, 71–72
TW white space, 108, 211
  service, 71–72
Typhoon Haiyan, 2013, 173
typhoons, 170

uAvionix, 126–27
Uccellini, Louis, 174–75
UK Met Office, 216
Understory, 123
Unified Forecast System (UFS), 180
United Kingdom, 19
United Nations, 211
United Nations Environment Programme, 68
United Nations Office for the Coordination of Humanitarian Affairs (OCHA), 227
United States, 19
  politicized environments of, 235
  southeastern, 35–37
University at Albany, 230–31
University Corporation for Atmospheric Research (UCAR), 92
University of California, Berkeley, 119
University of Oklahoma, 21
University of Reading, 228
University of Wisconsin-Madison, 52
unmanned aerial vehicle (UAV), 102
upwelling, 217
urban heat-island mapping campaign, 129
US Agency for International Development, 226
US Department of Commerce, 15, 105
US Department of Defense, 125
US government, 91
US National Hurricane Center (NHC), 4
US Occupational Safety and Health Administration (OSHA), 145
US Weather Bureau, 184

Vaisala, 28
Valley Fire of 2020, 49
Vallgren, Andreas, 112

van Oldenborgh, Geert Jan, 153–54
vasoconstriction, 141
vasodilation, 141
Vernon, John, 102–3, 126
Vines, Jennifer, 149
virtual scenes, 198
volunteers, 157
von Neumann, John, 184
VORTEX, 9, 35–37

Walker Building, 85
wall cloud, 13
warned event, 30
Warner, Charles Dudley, 1
warning fatigue, 33
warnings, 20
  encouraging attention to, 44
  ignored, 9, 159
  nighttime, 37
  nine minutes of, 38
  red-flag, 64
  ten to twenty minutes, 41
  time of day, 37
warning time, 31
warn-on-detection approach, 41
Warn-on-Forecast System (WoFS), 41–42
Warren, Earl, 65
Washington State, 154
water, 199
Waugh, Sean, 11–13, 20, 23, 45–46
wave (in the air), 169
Waymo, 122
wearables, 163
weather
  controlling it, 1
  defending ourselves from it, 1
  doing anything about it, 1
  doing something about, 9, 229–238
  extreme, 2
  future of, 237–38
Weather (Apple), 103
weather balloons, 46
Weather Channel, 6, 104, 162, 198–99, 201
Weather Channel app, 123
weather communicators, 6
Weather Company, 123
weather disasters, 3

weather enterprise, 105, 113
  people of the, 9
  triad of, 8
weather enterprise, primary groups in, 91–92
weather events
  destructive, in modern era, 1–2
  people killed and injured, 9
  stages of, 7
WeatherFlow-Tempest, 118
weather forecaster, target, 89
weather forecasting, 5–10
  by public and private organizations, 104–5
  comparisons of, 178
weather forecast offices (WFOs), 74, 79, 96–97
  local officials, 78
  products, cadence, and vocabulary of, 98
weather inequality, 6
weather literacy, 237
  skills, 6
weather meeting, 191
weather models, resolution of, 104
weather balloons, 12
weather people, as cloud warriors, 9
weather prediction, 2
Weather Prediction Center (WPC), 74, 147–48, 159
weather-prediction models, accuracy of, 159
weather predictions, accuracy of, 4
Weather Radio, 83–84
Weather-Ready Nation, 78
weather report, 73–100
Weather Research and Forecasting Innovation Act, 15
weather satellites, 186, 187
Weather Service, 73, 97

weather station, 123
  personal, 116
  home, 117
Weather Story, 81, 96
Weather Surveillance Radar, 1988 Doppler (WSR-88D), 23
Wellenius, Gregory, 140–41
Wellington, Alisha, 88
West Africa, 112
Wheeler, Jonathan, 125
Wichita, KS, 94
Wi-Fi, 108
Wildfire Analyst, 66–67
wildfires, 2, 48, 51–52
wildfire season, cost of, 53
wildland-urban interface (WUI), 49
  growth of, 52
William Penn Foundation, 135–36
wind, science of, 60
winds, 60–61
wind speed, 176
wireless signals, 114
Witch Creek Fire of 2007, 58
World Food Programme (WFP), 211, 226, 227
World Health Organization, 221
World Meteorological Organization (WMO), 2
World Weather Attribution, 153–54
Wunderlich, Carl, 141

year 2023, hottest recorded, 2
yes-or-no answers, 212

Zimbabwe, 210–11, 224–25, 227
  Meteorological Services Department, 225–26

About the Author

THOMAS E. WEBER is a veteran writer, editor, and newsroom executive with a lifelong interest in how science and technology shape society. He was *The Wall Street Journal*'s first internet columnist and later became a bureau chief for the paper. As executive editor of *Time*, he supervised the magazine's feature journalism. Tom has taught journalism and publishing at Columbia University and New York University as well as at his alma mater, Princeton University. He lives in New York City with his wife. Visit thomaseweber.com for more.